Advancing Space Physics

Auroral Substorms
and Solar Flares

Syun-Ichi Akasofu

AMERICAN ACADEMIC PRESS

AMERICAN ACADEMIC PRESS

By AMERICAN ACADEMIC PRESS

201 Main Street

Salt Lake City

UT 84111 USA

Email manu@AcademicPress.us

Visit us at http://www.AcademicPress.us

ISBN: 979-8-3370-8943-0

Distributed to the trade by National Book Network Suite 200, 4501 Forbes Boulevard, Lanham, MD 20706

10 9 8 7 6 5 4 3 2 1

To

My wife Emiko

Preface

I wrote this book for researchers who intend to make new advances in space physics and solar physics. I provide several suggestions on how to make advances based on my experiences.

I developed a new field of auroral research in 1964 — auroral substorms and magnetospheric substorms. This field has become one of the most studied subjects in space physics and has been considerably advanced. I describe how this study began and has developed, as well as the progress of space physics along the way from the earliest days of the 1960s.

The basis of my study in both magnetospheric physics and solar physics is the *electric current approach*, which was initiated by Hannes Alfven. In this approach, each phenomenon in space physics and solar physics is explained by following electric current, from dynamo (power supply), transmission (current/circuit) to dissipation (manifested by observed phenomena). This approach is quite different from the traditional approach, which considers magnetic field lines as the base, namely the magnetic field line approach. The differences are pointed out.

Readers will find that this electric current approach might provide a new insight into several major unsolved problems for decades, such as auroral substorms, solar flares, the solar corona and the solar wind. Thus, I hope that this book will serve as an introduction to the auroral physics and space physics in a unique way, the electric current approach.

All the chapters are mainly based on my own experience with concrete examples. This differentiates this book from any natural science textbooks and monographs (mostly explaining known facts and prevailing theories). I summarize my methodology in Chapter 8. Hopefully, young readers will be stimulated in this unique way. Thus, this book is not intended to be a review.

In Chapter 1, I describe the history of auroral and space science from the beginning (1845/1893) to the end of the 1950s, when the space age began. It will provide the needed background for young researchers, particularly for later chapters.

In Chapter 2, I describe first how I worked on an unsolved problem on geomagnetic storms, which Chapman and many others had worked for 30 years, but did not seem to succeed. In this work, I explain how I, as a beginning graduate student, found the *"unknown"* factor (unthinkable at that time) in the solar wind, which opened a new way of studying auroral physics and magnetospheric physics.

Chapters 3 describes my study of the aurora based on visual, all-sky cameras and satellite images. My simple (visual) observation of the aurora disagreed with the well-established auroral distribution, the auroral *zone*. I supported the new concept of the auroral distribution, the auroral *oval* established by Yasha Feldstein, by various methods.

Chapter 4 includes a story about how I developed a new branch of auroral physics; auroral dynamics, which has developed into *auroral and magnetospheric substorms*. This study began in satisfying my curiosity on auroral activities over the whole polar sky, but the results took many efforts to confirm and convince others. This effort might also serve as an example of how to convey a controversial result to many others.

In Chapter 5, I attempt to understand physical processes of auroral substorms as a large-scale *electrical discharge* in terms of a sequence of power supply (dynamo), its circuits and energy dissipation (observed phenomena). My electric current approach was suggested by Hannes Alfven, who initiated the field of magnetohydrodynamics (MHD) in 1950, but later (1967) emphasized *the electric current approach*. However, this approach requires how the intensity of electric current varies as a function of time during auroral substorms. In fact, the results show the importance the electric current study in Chapter 6.

In Chapter 6, I have established a morphological theory of auroral substorms, with my colleagues, by synthesizing various observed facts. It is exclusively based on the electric current approach. Thus, this book is unique in this respect, although the prevailing theories are based on the magnetic field line approach. In this chapter, I emphasize that many auroral and solar problems are very difficult to understand without considering the electric current approach; in particular, the development of the double layer in field-aligned current is crucial in understanding the whole processes of auroral substorms.

In Chapter 7, I extend my study of auroral substorms to unsolved solar physics problems, including the coronal ionization, the cause of the solar wind, solar flare and sunspots in terms of the electric current approach. The photospheric dynamo plays a crucial role in proving the necessary power. I try to show that the electric current approach might provide a new sight into these difficult problems.

In Chapter 8, I summarized the development of my own methodology. The first part explains it by concrete examples. Most of us are working under well-accepted theories, which are taught generation after generations for few decades. If one wants to make a stepwise advance in his/her field in this situation, what one might do ? I hope that this chapter will be useful for young researchers in establishing their career.

I would like to thank the late Sydney Chapman, the late Hannes Alfven and the late Walter Roberts for their guidance in my research life. Without them, my scientific career would not have existed. Taking this opportunity, I would like to thank also all of my colleagues, regardless of their agreements and disagreements with me. Without them, it would be impossible to write this book. They are listed in my second book titled "Physics of Magnetospheric Physics" (D. Reidel, Pub.) and "Solar-Terrestrial Physics", co-authored with Sydney Chapman (Oxford Univ. Press).

I would like to thank specially B. H. Ahn, Yasha I. Feldstein, Lou A. Frank, Y. Kamide, Carl E. McIlwain, Ching- I. Meng, Lou-Chang Lee, A. Tony Y. Lui, A. Lee Snyder, Paul Perrault, Sun Wei and Bruce Tsurutani for their very close collaboration in my research for a long time. I would like to thank Ned

Rozell, Robin Nicholson and Keiko Herrick for their help in preparing the manuscript.

Since this book is an introductory book on the electric current approach, both figures and photographs are intended to be visual aids for reading the text, not for detailed examinations. However, they are cited and referenced at the end of each chapter. The same figures appear a few times for easy of reading, because many important points appear in several chapters. Further, since most of the readers would not be familiar with the Electric current approach, the subject index is designed to find specific subjects they are looking for.

Since I did not write my diary, some dates are not given in order to avoid inaccuracy. The photographs and figures are assembled from what I happened to have during the last 60 years (not the purpose of collection or writing this book) and thus some of them are unfortunately not credited because my memory is uncertain; I am very sorry for this omission. Many figures provided by the Geophysical Institute, University of Alaska are noted by (GI).

The proposed cover photographs (Credit)

Top row: Solar corona (Y. Kozuka), Single sunspot (The Kitt Peak Solar Observatory), Solar flare (K. Shibata).

Middle row: Auroral photograph (L. Snyder), All-sky image of the aurora (GI), Auroral oval (L. Frank),

Bottom row: Coronal mass ejection (NASA), Auroral substorm (L. Frank), Aurora on Saturn (Boston Univ).

Contents

Chapter 1 Aurora and space physics in history

1.1 A few earliest history

"Aurora" is the name of the Roman goddess of dawn, but it is not accurately known how this name was introduced as a scientific term or who was the first person to use it as a scientific term. The term "northern lights" is very often spoken by northern people.

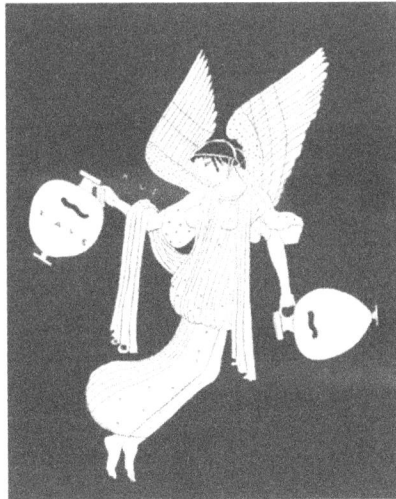

Figure 1.1 Aurora. Roman Goddess of dawn.

In Roman days, stars were thought to be holes in the sky. As for the aurora, Seneca mentioned: "There are chiasmata [fissures or canyons], when a certain portion of the sky opens, and gasping displays of flame as in a torch."

During the Middle Age, people sketched the aurora with their imaginations, such as a battle in the sky, when they saw swiftly moving 'rays' (vertical striations; see on them later in this section; Section 1.2) in the aurora or a series of candles in the sky.

Figure 1.2 Sketches of the aurora in the middle age (Left: Zentralbiothek Zurich: Right: Astronomer Royal of Scotland).

In his book "The Aurora Borealis," Alfred Angot (1896) described the terror caused by red auroras: "Pilgrimages were organized to avert the wrath of Heaven."

Figure 1.3 Red aurora described by Alfred Angot.

There are many legends and stories of the aurora among northern native peoples, including haunting ghosts, lightning snakes and the fox throwing up snow with its tail to create the aurora.

Figure 1.4 There are many legends and stories on the aurora among the northern native people.

During one of the many Arctic expeditions of the 19th century, the British Admiralty was anxious to find the Northwest Passage between the Atlantic and Pacific. This mission resulted in a great tragedy with the loss of Sir John Franklin, the leader and his crew. Their search parties spent many nights in the Arctic and witnessed the aurora. In one of their reports, Charles Hall (1865) described it:

"Who but God could conceive such infinite scenes of glory?"

There have been many stories on this tragedy even today.

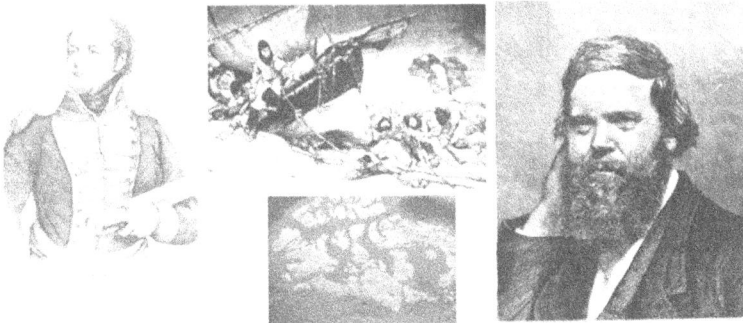

Figure 1.5 Left: Sir John Franklin and his tragic expedition in the Northwest Passage in about 1850; red dots are the location where their remains are found. Right: Charles Hall.

Later, some early Arctic explorers sketched the aurora accurately. Norwegian

Fridtjof Nansen was a good artist as well as a great explorer and sketched aurora accurately during his Arctic Ocean expedition. His woodcuts show clearly that the aurora has a curtain-like structure. (The curtain-like structure is not necessarily obvious even in many recent photographs taken by sensitive cameras. His woodcuts show also the ray structure (Section 1.2); the rays are more difficult photograph.

Figure 1.6 Left: Fridtjof Nansen. Right: His woodcut, showing the aurora and his ship Frum (The Fram Museum).

Figure 1.7 Left: A woodcut by Fridtjof Nansen. Right: a photograph at a similar situation (Takeshi Matsuo). Nansen was very accurate in sketching the aurora. His sketch shows clearly that the aurora has a curtain-like structure with the ray structure.

Harald Moltke, a famous Danish painter of the aurora, accurately sketched the aurora. When an active auroral curtains appeared right overhead (more accurately, the magnetic zenith, namely the point of converging point [upward extension of magnetic field lines at the observing point]), he must have memorized the scene, which lasted just a few seconds. This type of the appearance of the aurora was called 'corona' by considering it was a different type of the aurora.

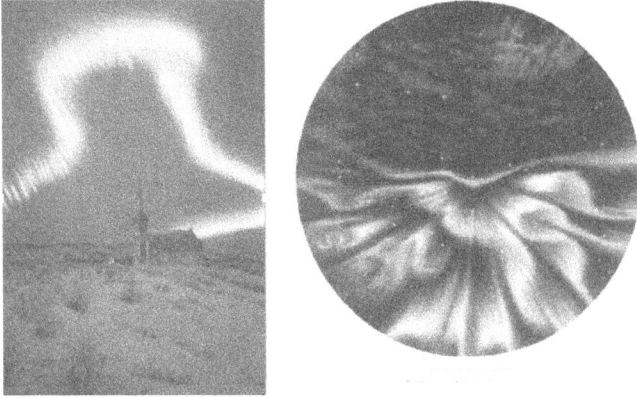

Figure 1.8 Two paintings by Harald Moltke. He could so accurately paint when an active auroral curtain is located right overhead. He must have memorized the scene (the corona type), since such a scene lasts for only a few second. A set of his paintings of the aurora was given to the author by the University of Oslo. They are now displayed at Poker Flat Rocket Range of the Geophysical Institute of the University of Alaska Fairbanks.

Many early Arctic visitors wrote books about what they saw on the aurora and described its beauty and majesty. Those books were widely read by lower-latitude people.

Figure 1.9 Books by early arctic travelers. Left: Sophus Tromholt's book (1885). Right: Easter Bardsall Darling's book (1885).

In early days, people sketched the aurora very accurately.

TRAVELLING IN LAPLAND.

Figure 1.10 Left: Sophus Tromholt's sketch of the aurora and a similar photograph; it shows how accurately he sketched the aurora.

1.2 Basic facts of the aurora

The basic form of the aurora is a curtain-shaped structure aligned along magnetic field lines (thus, not strictly vertically aligned). The thickness of the curtain is (in photographs) only about 500 meters; apparently, I was the first to measure the thickness by a camera (Akasofu,1961).

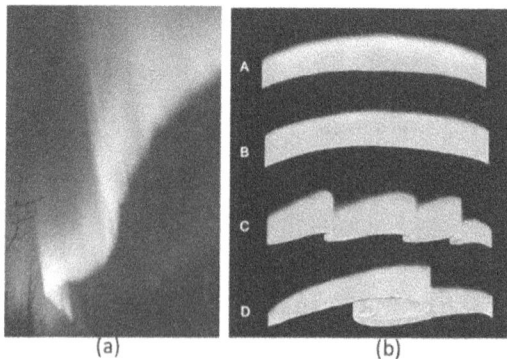

(a)   (b)

Figure 1.11 (a) The photograph shows a curtain-like shape of the aurora. (b) Auroral curtain develops various size of folds (including the ray structure) when it becomes active. The size of the folds becomes larger when the activity increases.

6

The curtain-shaped aurora is often called an "arc" because the auroral curtain appears often as arc or an arch in the northern horizon. The curtain structure appears very often in multiple; Figure 1.13.

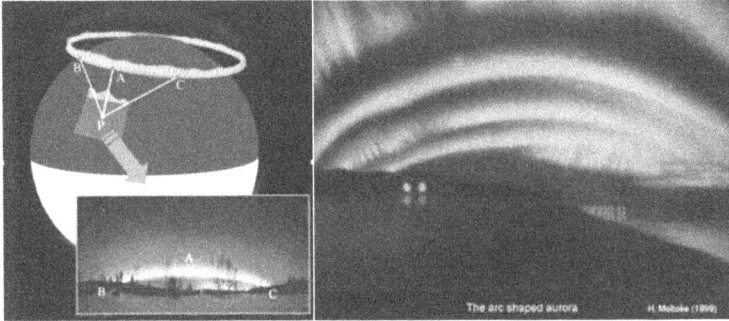

Figure 1.12 Left: The auroral curtain appears like an arc or arch in the northern sky. This is the reason why the curtain-like form of the aurora is called 'arc'. Right: Harald Moltke's painting.

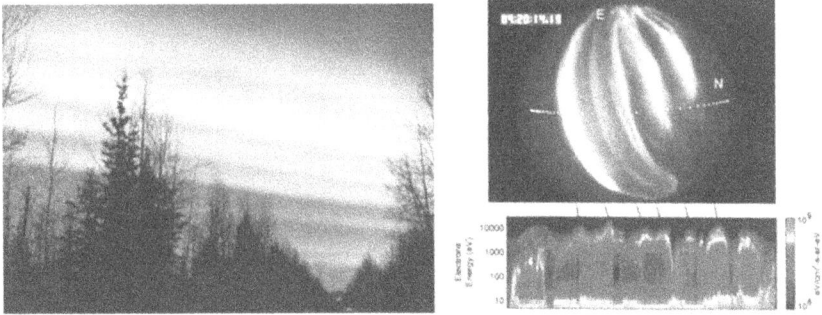

Figure 1.13 Left: Multiple appearance of auroral curtains (GI). Right: Upper; An all-sky image, showing multiple arcs (GI); Lower, Satellite observation of the precipitation of electrons in a multiple structure (GI), the so-called 'inverted V' structure. In terms of the electron precipitation, the width of the curtain is a few kilometers (photographically or visually about 500 m).

Since the aurora is a large-scale phenomenon covering the whole sky, it appears different, depending upon one's relative location with respect to the aurora. As the curtain stretches beyond the horizon, it appears as if it originated from the mountain top (so reported in an early Encyclopedia Britannica). Particularly, when it appears overhead, the aurora appears as light columns (the ray structure and folds) emanating from a point in the sky (the magnetic zenith as mentioned earlier. Moltke showed such a display in his painting, as shown in Figure 1.8 [right]). A large-scale loop is actually a looping of the curtain.

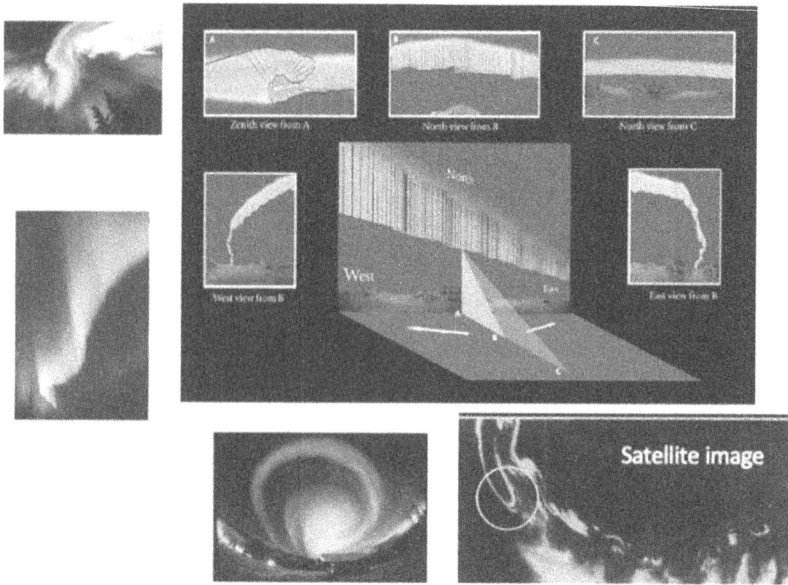

Figure 1.14 The auroral curtains appear to look different, depending on the relative location from observers. In some cases, the aurora appears like a large-scale loop (GI), but a satellite image shows a folded auroral curtain (DMSP).

When the auroral curtain is very quiet, the brightness is horizontally fairly uniform, and shows the vertical (actually along the inclined earth's magnetic field lines) ray-like structure when it becomes active. It is actually the smallest folds (pleating) of the auroral curtains that appear as rays in the curtain. The ray structure has the length of about 400 km (vertical width of the curtain) and is clearly shown as small folds by a high-resolution and high-speed TV camera by looking up the bottom of the curtain; a computer simulation (Wagner et al., 1985) shows that the rays are caused by a fast counter-flow of ionospheric plasma along the curtain associated the double layer (Section 6.4).

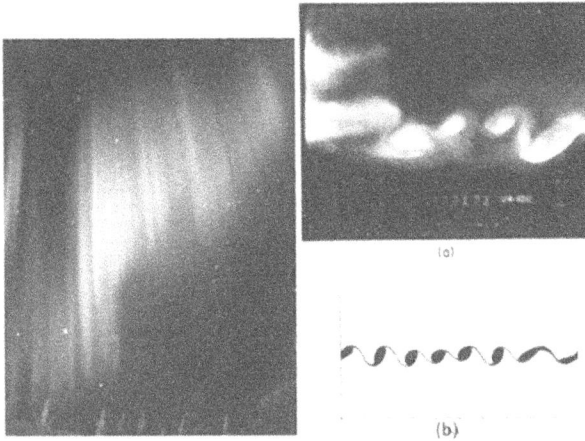

Figure 1.15 Left: The auroral curtain shows the ray structure, called 'rays' (GI). Right: (a) The rays are actuarily small folds of the auroral curtain and is clearly shown by a high resolution and high speed TV camera when looking up the bottom of the curtain (Tom Harinan). (b) A computer simulation of rays by assuming a counter flow (along the curtain) of ionospheric plasma (Wagner et al. 1985). The counter flow is caused by the double layer (Section 6.4).

In the early 1900s, Norwegian scientist Carl Stormer was the first person who determined accurately the bottom height of the aurora by triangular observation; they got help from a telephone company. The bottom height of the aurora is about 100 km (about 60 miles) above the ground. It is about 10 times higher than the flying altitude of commercial airliners. The upper height is about 400 km, but often extends up to 1,000 km when it is very active (its upper part then often becomes red in such cases). The aurora is seen often as red light from lower latitudes in the northern sky, because they can see only the upper part of the curtain near the northern horizon (often confused as a fire in a northern town in the past). Space Shuttles fly above the brightest part of the aurora. (100-120 km high).

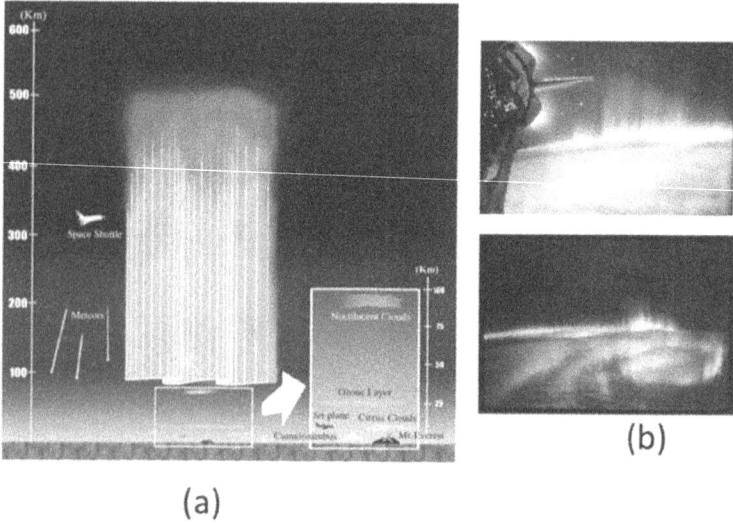

Figure 1.16 (a) Height of the auroral curtain; the lower atmosphere is also shown. Note the different colors at different altitude (see the spectrum in Figure 1.18). (b) The aurora from Space Shuttles (NASA).

Figure 1.17 Hokkaido Newspaper reporting the red aurora; October 23, 1989 (Hokkaido News Paper).

However, during great geomagnetic storms, the aurora can be seen not only in midlatitude, but even in Arizona, Mexico and Hawaii (Section 3.4); it appears as structured or diffuse glow.

If we view the aurora through a prism, auroral light shows itself quite different from sunlight (a rainbow of light, continuous from red to violet). The auroral light consists of only a few lines and bands. Similar line lights can be seen in a neon sign, which is a thin vacuum tube (containing a small amount of neon gas) connected to a high voltage supply, namely a *high voltage electrical discharge in a vacuum*. The degree of vacuum at auroral heights is much higher than that of a neon sign. This is the subject of Section 5.1.

11

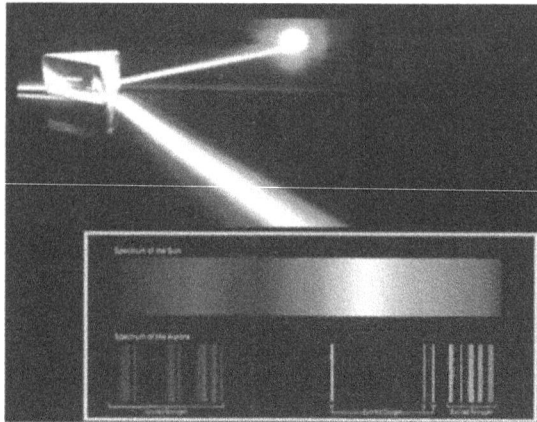

Figure 1.18 Comparison of sunlight (upper) and auroral light (lower) by looking through a prism. In Figure 1.16, the pinkish emission near the bottom of arc is the band about 66,00 mm from N-O molecules, the middle green light is 55.77 mm line emission from oxygen atoms and the red light is from oxygen atom. See Figure 1.19.

The most common light of the aurora is greenish, a color emitted by atomic oxygen. In the upper atmosphere, oxygen molecules are split into two atoms by solar X-rays. They emit the green light when they collide hardly with electrons (4 eV). However, when they collide softly (2 eV), they emit the red light. These electrons are secondary electrons produced by the primary collision of very energetic electrons (10 KeV). A third light is pinkish, existing near the lower bottom of the auroral curtain (about 90 km in altitude). These are emitted by nitrogen-based molecules; for details of auroral spectrum, see Chamberlain (1961).

(a)   (b)   (c)

Figure 1.19 (a): Greenish light from oxygen atoms. (b) Pinkish light from nitrogen molecular compound. (c) The red aurora on February 11, 1958 (GI)

1.3 Early theories

Anders Celsius (1701-1744), professor at Uppsala University, Sweden, was one of the first scientists to recognize that the tip of a long magnetic compass slightly vibrated during auroral activities.

Based on extensive auroral observations, Edmond Halley (1656-1742), the discoverer of the Halley Comet, considered that some particles run along earth's magnetic field lines, penetrating the earth. Halley was aware at least some *particles* and *geomagnetic field lines* are related to the aurora (since the earth's magnetic field was reasonably known for ocean sailings at that time).

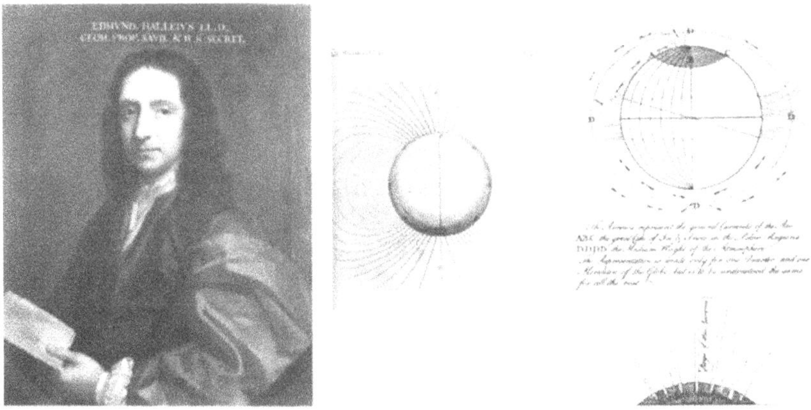

Figure 1.20 Left: Edmond Hallett and his idea on the cause of the aurora; he thought that particles moving along geomagnetic field lines caused the aurora. Right: Benjamin Franklin 's idea on the cause of the aurora.

Benjamin Franklin (1706-1790) postulated that moisture carried by wind from the tropics falls in the polar region as snow, and that an *electrical discharge* takes place between the snow and air. It was prescient of him to consider an electrical discharge as the cause for the aurora; he learned science in England.

In 1904, Selim Lemstrom, at the Sodankyla magnetic observatory in Finland, tried to reproduce the aurora with a series of discharge tubes, but it is not certain if it worked (because of incomplete circuit). However, it was a very advanced idea compared with many earlier ones, because he used a series of *vacuumed electrical discharge (neon sign) tubes* for the aurora and a high voltage supply. He made also his own observation at Spitzbergen (now, Svalbard), and reported once he was within the aurora. This is understandable; if one observes very active aurora alone at the top of hill, it is possible to get overwhelmed by active auroras.

In fact, in Section 5.1, we discuss the aurora as a majestic electrical discharge around the earth and use a neon sign to explain the auroral electrical discharge.

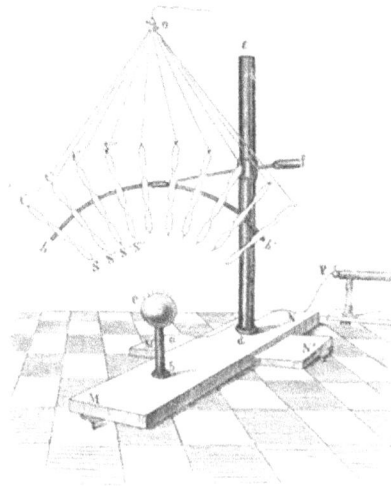

Figure 1.21 Selim Lemstrom's experiment in 1904 (Sodankyla Observatory). Note a series of discharge tubes.

1.4 Dawn of space physics

(a) Discovery of solar flare

The history of space physics had a very rocky start. The discovery of a solar flare, an eruption on the sun, was made by R. C. Carrington, a British solar observer, at the Greenwich Observatory; (Carrington, 1860).

When he was making a routine sketch of sunspots, there appeared suddenly bright spots around a group of sunspots only for a short time on September 1, 1849. In those days, automatic photography was available (perhaps, only few times a day), but fortunately he refused to use it and was making sketches.

He observed a white light flare (visible), the most intense type of solar flare. On the next day, an intense geomagnetic storm occurred on the earth recorded by magnetometers at several places, including the Greenwich and India. There occurred also troubles in electric-wire communication systems (produced by induction currents on the ground). Solar flares are the subject of Section 7.1.

The aurora was seen in many places on the same night, including Hawaii and Mexico. Carrington reported his finding and the subsequent occurrence of the geomagnetic storm and the aurora by stating: "One swallow does not make a summer." He himself doubted the coincidence of the two phenomena.

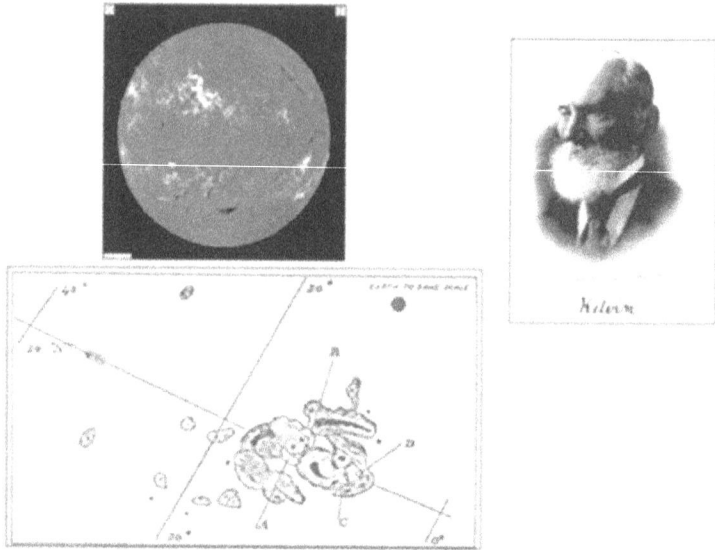

Figure 1.22 Left: An example of solar flares. Carrington's sketch of the first sighting of solar flare (Royal Astronomical Society, London). Right: Lord Kelvin (American Geophysical union).

The occurrence of the solar flare and geomagnetic storm together with Carrington's modest statement got the attention of Lord Kelvin (1824-1907), the great physics authority in those days. In his Royal Society Presidential Address (Kelvin, 1892), he claimed that the two observed facts were "a mere coincidence." (Obviously, he did not know that the sun ejects a cloud of ionized gases, which are now called coronal mass ejections [CMEs] or magnetic clouds [MCs]). Since he referred to Maxwell's theory in his speech, he must have thought that magnetic changes on the sun could not be detected at the distance of the earth. Solar flares, CMEs (or MCs) are the subject of Section 7.1 (h).

However, it was in 1905 when Edward W. Maunder (1905) at the Greenwich Observatory, stated that the sun is responsible for geomagnetic storms on the basis of the recurrence tendency of geomagnetic storms (of 27 days), which coincides with the rotation period of the sun (seen from the earth). This was against the strong objection mentioned by Lord Kelvin.

However, Maunder was convinced that geomagnetic storms are caused by a "stream" of gas from the sun.

Maunder stated (Maunder, 1905):

"First: The origin of our magnetic disturbances lies in the sun---."

This was the very *first* statement in the history of space physics *based on the observation* that the sun is responsible for the occurrence of geomagnetic storms. However, there were objections to his statement. A prominent physicist in those days stated: "The mystery is left more mysterious." What we have thought to be a "stream" from "coronal hole" is the subject of Section 7.3(d and e).

Figure 1.23 Edward W. Maunder. His name is known under "Maunder minimum" (1645-1715). It was the period when sunspot number was very low, and this period was named so by recognizing his contribution (Royal Astronomical Society London).

(b) Birkeland's electron beam theory

After electrons were discovered by Joseph John Thomson in 1887, Norwegian auroral scientists Kristian Birkeland and Carl Stormer believed that the sun emitted a beam of electrons, which caused both magnetic disturbances and the aurora. The electron theory by Birkeland and Stormer had prevailed for a few decades (1920-1950); Birkeland (1918). They were the pioneers in the field of auroral physics and space physics.

Birkeland set up a large vacuum box, in which he placed a magnetized iron ball representing the earth and shot an electron beam to it in order to reproduce the aurora on the painted iron ball. The vacuum box is now maintained at the University of Tromso. His device still works!

Electron beam from the sun

Birkeland's book

K. Birkeland
Norwegian physicist

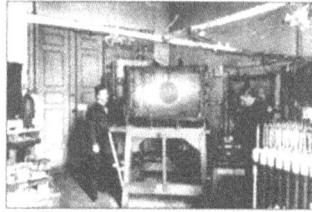

Tromso University, Norway

Figure 1.24 Left: Kristian Birkeland (the University of Oslo). Middle: Birkeland's experiment, in which an electron beam was shot to toward a magnetized sphere in a large vacuum box. This box is now at the Tromso University, and it still works! Right: Birkeland's book on his arctic expedition.

Inspired by Birkeland, Stormer began his life-long study of motions of electrons around the earth's dipole field. This was a very extensive numerical work in those days. As mentioned earlier, he was also an active observer of the aurora and determined the height of the aurora accurately for the first time. He published a book *"The Polar Aurora"* (Stormer, 1955). It is interesting to note that his study of charged particles in a dipole field was very useful for James Van Allen, who discovered the Van Allen radiation belts in 1959, and also for my study of the ring current (Section 2.3).

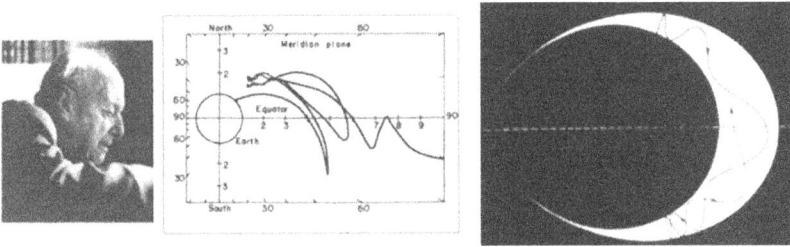

Figure 1.25 Left: Carl Stormer. Middle: An example of the trajectories of electrons from the sun. Middle: Trajectory of a charged particle from the sun. Right: a trajectory of a particle trapped in a dipole field (University of Oslo). The particles in the Van Allen radiation belt and in the ring current follow such a trajectory (Section 2.3).

Figure 1.26 Photographing the aurora. The sitting one is Carl Stormer (the University of Oslo).

Lars Vegard was stimulated by Birkeland and became the first person to study the spectrum of the aurora. He found the hydrogen emission on the spectrum. The history of studies of auroral emission has a long history (cf. Chamberlain, 1961).

19

Figure 1.27 Larus Vegard was the first auroral spectroscopist (University of Oslo).

(c) Chapman's plasma flow theory

Sydney Chapman was employed at the Greenwich Observatory, and the observatory director showed him Maunder's work. Chapman (1918) decided to theorize how Maunder's stream from the sun can produce geomagnetic storms; it became one of his life-long works. Chapter 2 describes his work in detail.

Chapman told me that he also first considered a beam of electrons (or protons), but Lindemann (1919) criticized Chapman's theory by saying that an electron beam from the sun would not reach the earth because its electrostatic repulsion. Lindemann thought that the temperature of the sun was about 6000° K and "ionization would be almost complete". Thus, he suggested that if the gas is consisted of protons and electrons (what we now call *plasma*), it might be able to reach the earth.

Therefore, Chapman and his graduate student Vincenzo Consolato Antonio (V. C. A.) Ferraro began to work on this problem. They first made sure that the solar stream in the vicinity of the earth's dipole field could be treated as plasma by inventing a formula (similar to the Debye length); they found that its frontal surface of the flow can be treated like a perfect conductor. They published their paper, titled: *"A new theory of magnetic storms"*, in Terrestrial Magnetism and Electricity, which has become Journal of Geophysical

Research later; (Chapman and Ferraro, 1931). In it, they predicted a comet-shaped structure around the earth in the solar gas flow, which is now called the *magnetosphere.*

Chapman told me later that he was afraid that the problem was a very hard one for Ferraro, a graduate student. After almost 10 years of effort, they completed their theory. Ferraro mentioned their joy of accomplishment in *"Chapman Eighty, from his friends."* (University of Colorado Press). I visited Ferraro once at the University of London.

Their work has become the foundation of magnetospheric physics and space physics. *It is also the paper that inspired me to work with Chapman in Alaska;* Section 2.2. Details of the joint works with Chapman is the subject of Chapter 2.

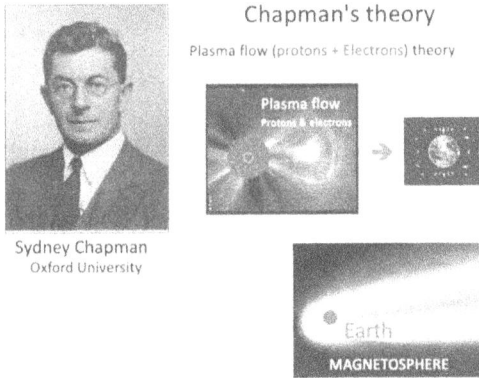

Figure 1.28 Left: Sydney Chapman (The Royal Astronomical Society of London). Right: Chapman-Ferraro theory. The solar plasma flow (ejected from solar activities, NASA) establishes a comet-shaped space around the earth' magnetic (dipole) field.

Chapman and Julius Bartels published a joint book *"Geomagnetism, Vol. I & II)"* in 1940. It was a great reference book in geomagnetism for a long time. Chapman told me that it was very difficult to work with Bartels in Germany at that time (just before the WW II).

Figure 1.29 *Geomagnetism* by Sydney Chapman and Julius Bartels and their photographs.

I had a great and fortunate opportunity to work with Chapman between 1959 and 1972 (the year of his death). His life is described in Section 2.3.

(d) Alfven's theory of geomagnetic storms and the aurora

Alfven (1950) published one of the most influential books in space physics, solar physics and astrophysics, titled *"Cosmical Electrodynamics."* In it, he established the theory of magnetohydrodynamics (MHD). In his theory, he introduced the concept of *"frozen-in"* magnetic field lines, in which a group of plasma particles on a particular magnetic field line remains with it. This concept has been extensively used in space physics and astrophysics. In this book, a study based on this concept is called *the magnetic field line approach.*

Figure 1.30 Hannes Alfven (H. Alfven).

On the other hand, in as early as 1968, in his paper, *"The second approach in cosmical electrodynamics"*, Alfven emphasized the need for the *electric current approach* in space plasma physics by emphasizing the importance of *electric current*. He warned: [if we neglect electric current], "we deprive ourselves of the possibility of understanding some of the most important phenomena in cosmic plasma physics." Alfven (1981, 1986) repeated this point in his second book titled *"Cosmic Plasma"*.

In particular, Alfven !1968) noted:

"One has good reasons to suggest that there often exist electric fields with components parallel to the magnetic field. The existence of such fields may invalidate the 'frozen-in' picture in many cases. We may say that the first new principle is associated with a 'thaw' of the frozen-in field lines." This subject is described as the double layer (Section 6.4).

In fact, in Chapters 6 and 7, I will show the condition of "frozen-in" field breaks down at very crucial points in the process of auroral substorms, so that it is not possible to explain auroral substorms and solar flares by MHD; the double layer cannot be formed under the "frozen-in" condition, so that there would neither be the auroral substorms nor the solar flares without the double layer.

Thus, he recognized in as early as 1968 that what we can learn by MHD is limited, and warned that the electrical current approach is needed for unsolved problems.

Unfortunately, the space physics and solar physics communities have almost ignored Alfven's point and taken almost exclusively the magnetic field line approach.

The electric current approach considers space physics phenomena systematically in terms of electric current in an electric circuit, which consists of power supply (dynamo), transmission (currents/circuits) and dissipation (mostly observed phenomena). We can learn this approach in a concrete way in studying auroral substorms in Chapters 5, 6 and 7.

In this book, I take almost exclusively the electric current approach, instead of the MHD-based magnetic field line approach. Since the magnetic field line approach has long been adopted almost exclusively in the past, this is thus the major difference between this book and most other books in space physics. Thus, we take practically a *new approach in space physics.* Chapters 5, 6 and 7 describe auroral substorms and solar flares based on the electric current approach.

I met Hannes Alfven for the first time during "*The Birkeland Symposium on Aurora and Magnetic Storms*" in Sandefjord, Norway, on September 18-22, 1967. I recall that he emphasized to me personally the need for the electric current approach at that time. I visited him in Stockholm several times and also invited him in Alaska (Chapter 6).

Alfven had a long debate with Chapman on geomagnetic storms and the aurora. Chapman's plasma was not a magnetized one, while Alfven considered magnetized plasma. Alfven proposed his theory of the aurora and geomagnetic storms, in which there would be two neutral points (one in the morning side and the other in the evening side) near the earth by the linkage of solar magnetic field lines with the earth's magnetic field. There would be an *electrical discharge* between the two neutral points; the discharge currents produce geomagnetic storms and the aurora. We enjoyed the debates between Chapman and Alfven in many international conferences.

By looking back now, I realize (in some respect) that I have tried to combine their efforts. Chapter 6 is my product of such an effort, in which Alfven's electrical discharge occurs within Chapman's comet-shaped magnetosphere; see also Episode in Chapter 6.

(e) Biermann's theory of comet tail and Parker's solar wind theory

Interplanetary space had been thought to be vacuum until about the end of the 1950s, except for the occasional presence of solar gas (plasma) clouds and streams. Ludwig Biermann suggested (on the basis of his study of comet tails) that the sun is blowing out its atmosphere (the corona) in all directions all the time (Biermann, 1951). I had an opportunity to meet Biermann once at the Los Alamos Laboratory in New Mexico and discussed his observation.

Chapman told me that in his effort of explaining the zodiacal light and a high temperature of the upper ionosphere, he considered once that the solar corona might extend beyond the distance of the earth, but his solution had the remaining pressure at the outer boundary.

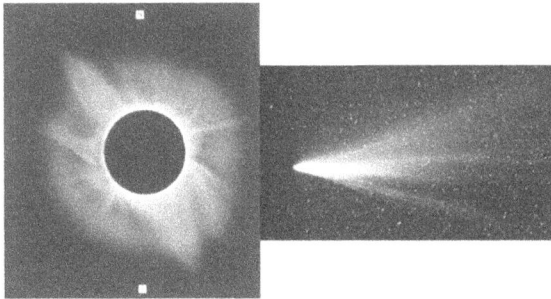

Figure 1.31 Left: The solar corona (Y. Kozuka). Right: A photograph of comet.

In 1958, Eugene (Gene) N. Parker (1958) theorized the flow of solar corona in terms of thermodynamic force and coined the term *solar wind*. It is my understanding that Gene understood Chapman's solution had remaining pressure at the outer boundary, which became the basis of his study of the solar wind.

The solar wind was detected for the first time by the Mariner-2 satellite in 1962. Thus, it has become certain that the sun blows out its corona all the time, and the comet-shaped magnetosphere considered by Chapman and Ferraro is a permanent feature of the earth.

Although Mariner 2 detected the solar wind, and this observation was considered to be the proof of Parker's theory. However, Joseph W. Chamberlain (1961) objected to Parker's theory on the ground that the thermodynamic condition of the solar atmosphere would not allow the observed high speed as Parker suggested. He called the plasma flow "solar breeze."

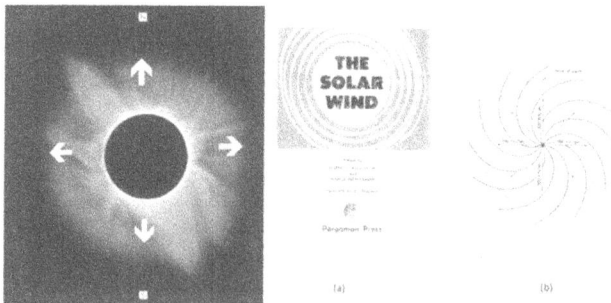

Figure 1.32 Left: (a) Schematic illustration of the blowing solar wind. Corona photograph (Y. Kozuka). Middle: The book on the First International Solar Wind Conference (published in 1966 by JPL). Right: The spiral interplanetary magnetic field lines (the stretched out solar magnetic field by the solar wind) proposed by Parker (1958).

In fact, the generation of the solar wind is still an unsolved problem; Lee and Akasofu (2021) suggested that a powerful Lorentz force (J x B), resulting from the solar unipolar induction system, is needed to overcome the solar gravitational force, but the (J x B) force could not reach the required speed. This is the subject of Section 7.3.

(f) Discovery of the radiation belts by Van Allen

The first scientific satellite launched by the U.S. allowed researchers to discover intense belts of radiation around the earth, the Van Allen radiation belts (Van Allen and Frank,1959). This finding played a very important role in the early development of space physics by stimulating space physicists. Van Allen told me that he had several rocket launches to study auroral particles in the Greenland area before his discovery of the radiation belts and found unexpectedly energetic electrons (perhaps, auroral electrons). Van Allen is mentioned more in Section 2.3.

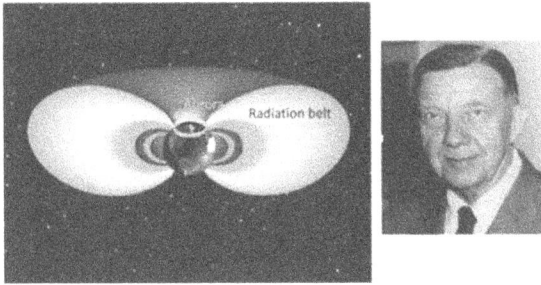

Figure 1.33 Left: Schematic illustration of the Van Allen radiation belts and the aurora. Right: James Van Allen (University of Iowa); See Section 3.3.

1.5 International Polar Years

(a) First International Polar Year

Among many polar explorations, the Polar Years were internationally most extensive. Karl Weyprecht conducted a polar expedition and was fascinated by the aurora and suggested an international observation of the aurora. The First Polar Year was operated in 1882-1883 after his death.

Figure 1.34 Karl Weyprecht

(b) Second International Polar Year

The second polar year was organized by many countries. Sydney Chapman and Julius Bartels were members of the organizing committee. It was held in 1932-1933, 50 years after the first one.

Figure 1.35 The committee member of the Second Polar Year.
Chapman was a member (S. Chapman).

Among many observers, Carlheim Gyllenskold made his auroral observation at Svalbard. He made a large number of all-sky sketches. There were also many other geophysical observational efforts during the Second Polar Year.

All-sky sketch

Figure 1.36 Left: Carlheim Gyllenskold. Right: His sketch of the aurora; he made many sketches in this format.

Adolphus Greely, an arctic explorer, participated in the Second Polar Year and reported about the aurora in his *"Three years of Arctic Service."*

"The aurora of January 21st was wonderful beyond description, and I have no words in which to convey any adequate idea of beauty and splendor of the scene. It was a continuous change from arch to streamers, from streamers to patches and ribbons, and back again to arches, which covered the entire heavens for part of the time. It lasted for about twenty-two hours, during which at no moment was the phenomena other than vivid and remarkable. At one time there were three perfect arches, which spanned the southwestern sky from horizon to horizon. The most striking and exact simile, perhaps, would be to liken it to a conflagration of surrounding forests as seen at night from clear or open space in their center."

(c) International Geophysical Year (IGY)

The International Geophysical Year (IGY, 1957-1958) was the third Polar Year, about 50 years after the Second Polar Year.

When Van Allen invited Chapman, Lloyd Berkner and a few others to his home, the idea of the Third Polar Year came up, which developed into the IGY. The men decided that geophysical fields other than auroral science should also be included. The fields of atmospheric sciences, seismology, volcanology and glaciology and others were parts of the IGY. The opening ceremony was held

29

in Moscow in 1957. Chapman was the president of the IGY.

Figure 1.37 The opening ceremony of the IGY in Moscow (1957). Chapman was the President (S. Chapman).

One of the main purposes of the IGY was to assemble geophysical data, and thus the World Data Centers (WDCs) was established in every subject. Chapman and Christian T. Elvey, director of the Geophysical Institute at that time, organized auroral observations during the IGY.

Scientists from many northern countries participated in this auroral observational project, devising their own cameras (16 mm or 35 mm size film) to record auroral activity over the whole sky, called the all-sky camera. They all agreed to take a photograph every minute during the dark hours. An example all-sky camera devised by the Geophysical Institute of the University of Alaska Fairbanks (GI) is shown in Figure 1.38. All the films thus produced at all the IGY auroral stations were assembled at the GI (WDC-Aurora).

I owe greatly both Chapman and Elvey for their efforts in establishing the IGY all-sky camera network. Research work throughout my research career has been based on the all-sky camera data. Most all-sky camera data used in this book are provided by WDC-Aurora at the Geophysical Institute, University of Alaska Fairbanks, noted by (GI).

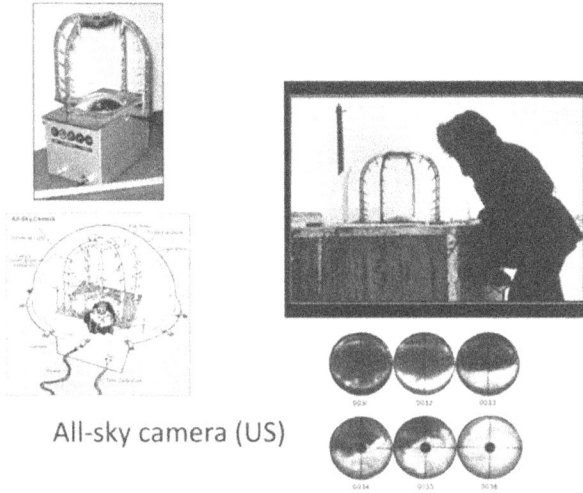

All-sky camera (US)

Figure 1.38 Left: All-sky camera and its structure (GI), Right: Daily checking of an all-sky camera (C. Deehr) and examples of the circular image of all-sky camera (GI).

Episode

I am fortunate enough to participate in a study of space physics from early days and further have known personally several pioneers, such as, Hannes Alfven, Sydney Chapman, Jim Dungey, Gene Parker; and James Van Allen. Readers will learn how they were through my association with them in this book. In fact, I owe greatly Chapman for his early guidance (1959-1972). It was also very important for me to have a chance to talk to early auroral scientists, such as Leiv Harang (Norway), A. Lebedinsky (Russia) and Jim Heppner (US).

Chapman told me that he saw Birkeland who was working his laboratory, but did not have the chance to talk to him. Stormer and Chapman had a great debate during a conference (Figure 2.28).

References

Akasofu, S.-I., 1961, Thickness of an active auroral curtain, J. Atmos. Terr. Phys., **21**, 287.

Akasofu, S.-I., Fogle, B. and B. Haurwitz, B., 1968, *Sydney Chapman Eighty*, Published by the National Center for Atmospheric Research and the Publishing Service of the University of Colorado.

Alfven, H., 1950, *Cosmical Electrodynamics*, Oxford Univ. Press.

Alfven. H., 1968, The second approach to cosmical electrodynamics,in *The Birkeland Symposium on Autora and Magnetic storm,* ed. byA. Egeland and J. Holt, Centre National de la Recherche Scientifique, Paris, 439.

Alfven, H., 1981, *Cosmic Plasma*, D. Reidel Pub. Co. Dordrecht-Holland.

Angot, A., 1896, T*he Aurora,* R. Paul, Trench,Trubner & Co.

Biermann, L., 1951, Kometenscheite undsolar korpsskularhlung, Z. Astrophys. **29**, 274.

Birkeland, K. R.,1918, *The Norwegian Auroral Polaris Expedition*, 1902-1903,

Carrington, R. C., 1860, Description of a singular appearance seen in the sun on September 1, 1859, Mon. Not. Royal Astrom, **20**,13.

Chamberlain, J., 1961, Physics of the aurora and Airglow, New York Academic Press.

Chapman, S., 1918, An outline of a theory magnetic storms, Proc. Roy. Soc. **97**, 61.

Chapman, S. and Ferraro, V. C. A. ,1931, A new theory of magnetic storms, Terr. Mag. Atmos. Elect.,**40**(4) 349.

Chapman, S. and Bartels, J., 1950, *Gomagnetism*, I & II, Oxford Univ. Press.

Easter Bardsall Daring, 1885, *Up in Alaska*, J. Anderson Press, Sacramento, California.

Hall, C., 1865, *Arctic researches, and lIfe among Esquimaux,* Harper, New York.

Maunder, E. W.,1905, Magnetic disturbances and their association with sunspots, Mon. Not. Roy. Astron. Soc. London, **65**, 2.

Lindemann, F. A., 1919, Note on the theory of magnetic stormsPhil. Mag., **38**, 669.

Parker, E. N., 1958, Dynamics of the interplanetary gas and magnetic field, ApJ., **128**, 664, https://doi.org/10.1086/146579.

Stormer, C., 1955, *The Polar Aurora*, Oxford University Press.

Tromholt, Sophus, 1885, *Under the Rays of the Aurora Brealis*, Mifflin, Boston.

Van Allen, J. A. and Frank, L. A., 1959, Radiation around the earth to a distance of 107,400 km, Nature, **183**(4659), 430-434, https://doi.org/10.1038/183430a0.

Wagner, J. S. et al., 1983, Small-scale auroral arc deformation, J. Geophy. Res., **88,** 8,013

Chapter 2 Geomagnetic storms: Ring current

Space physics originated historically in a study of geomagnetic storms (cf. Chapman and Bartels, 1940). In this chapter, we are mainly concerned with geomagnetic storms, particularly their main phase. The main phase is caused by the ring current around the earth. We learn what kind of motions of charged particles in the earth's dipole field produce the ring current and also what charged particles are in the ring current (including their origin). We find also that auroral activity is crucial in forming the ring current, which will be described in detail later in Section 6.7). This particular subject describes how I began to study both geomagnetic storms and auroral activities.

2.1 Beginning

The earth is a gigantic magnet. This was claimed by William Gilbert in 1600 in his book "De Magnete." However, its intensity and direction change a little (at most a few percent in the intensity and direction in a few minutes to decades). Thus, magnetic observatories have been set up internationally to observe these changes at many places in the world.

Figure 2.1 Left: William Gilbert claimed that the earth is a gigantic magnet in his book "De Magnete" (Dover edition). A magnetized sphere in iron powder, revealing magnetic field lines. Bottom: The earth with a compass needle.

The Tohoku University, my Alm Mata, had such an observatory at Onagawa near Sendai, Japan; Observatory was operated by professor Y. Kato, and all the instruments were located in a cave.

During my early undergraduate student days of the Tohoku University of Japan, I found a part-time job at the Onagawa observatory.

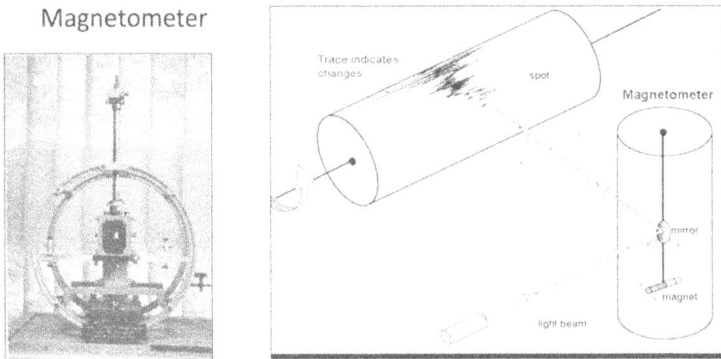

Figure 2.2 Left: One of the magnetometers at the Onagawa magnetic Observatory. Right: Inner working of a magnetometer. A beam of light is sent to a suspended mirror and magnet. The reflected light shows as a spot on the recording photographic paper.

My daily job was to change recording photographic papers of the instruments, magnetometers, in a dark room and develop the papers; magnetic changes appeared as a moving spotlight on a photographic paper wrapped around a cylinder that rotated once a day. The only thing I could see was a moving spotlight on the recording paper in the dark room in the cave. I was fascinated by delicate movements of the spotlight on the recording paper, but I knew nothing about the recorded traces on the paper.

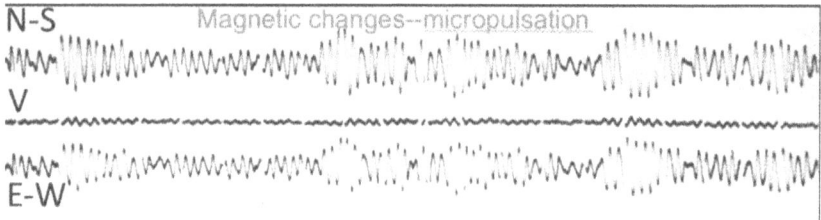

Figure 2.3 An example of magnetic record. It shows micropulsation (N-S, Vertical, E-W) components. Russian space physicist Valery A. Troitskaya called it "pearl necklace".

The observatory manager, J. Osaka, told me that some of large movements of the spot are caused by the aurora in Alaska or Siberia and that the cause of the aurora was not well understood.

This was the second time I heard the word "aurora". His word "aurora" reminded me my mother's song. I remembered during my earliest days that my mother used to sing her favorite (rather melancholic) song like a lullaby. It was called the "Wonderer's Song": "I am wondering whether I should go further ahead or go back home under the aurora." is one line of the song at the beginning.

I wondered about the '*magic hand*' (not knowing magnetic field related to the aurora), which could move the spotlight on the recording paper from such a great distance in Japan from Alaska.

Since I was deeply attracted by the mysterious movements of the spotlight, I decided to study the aurora and magnetic changes at that moment. Obviously, I knew nothing about the aurora and magnetism at that time. After all, this problem has become my life work, and I am still working on this problem.

2.2 Chapman's paper

Soon after the graduation, I had a job at Nagasaki for two years and wanted to study the aurora after returning back to my old school as a graduate student. However, I was not sure what exactly I could do to study the aurora at that time. On one occasion in a national meeting on the ionosphere, Professor Takeshi Nagata of the Tokyo University, asked me if I read an important scientific paper by Chapman (Chapman and Ferraro, 1931). He added, "Unless you can understand his paper, you are not qualified to study the aurora and magnetic changes [geomagnetic storms]."

After finding Chapman's paper in the library of my university, I found that it was very difficult to comprehend.

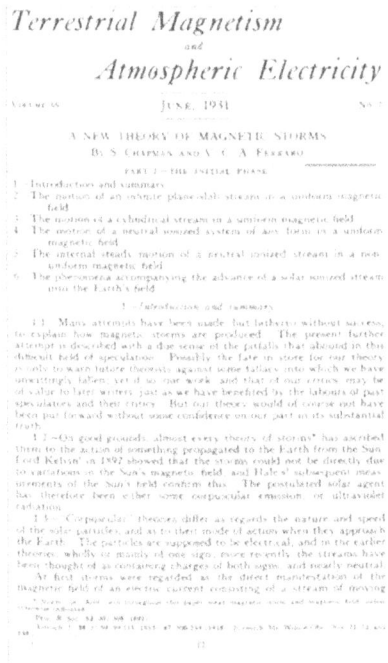

Figure 2.4 The first page of the paper by S. Chapman and V. C. A. Ferraro (1931) on their theory of geomagnetic storms. This was the paper which has determined my research life.

However, since I decided to study the aurora, I prepared a letter to the author,

Sydney Chapman, listing more than ten questions about his paper.

Then, I learned that Chapman was at the Oxford University (one of the top universities in the world) and the greatest authority in this field. Thus, I hesitated to send the prepared letter (from a beginning graduate student in Japan).

Then, a year or two later, I learned that Chapman had retired from the Oxford University and spends a few months every year at the Geophysical Institute of the University of Alaska (Fairbanks).

I was also told by my colleague of radio physics that the Geophysical Institute is the only institution in the world that is specialized in auroral research (at that time).

Thus, I decided to send my letter to Chapman in Alaska in April, 1958. I did not expect his reply to a poorly written letter from a beginning graduate student.

To my great surprise, just a few weeks later, I received his reply and a check (for travel), saying that I might study those questions under him in Alaska. I immediately accepted the offer, hoping also to have a chance to climb some Alaskan mountains. In any case, *his paper decided my destiny.*

Before leaving for Alaska, Professor Kato had a task that required me to participate in the total eclipse at an uninhabited coral reef (atoll) in the South Pacific Ocean on October 12, 1958.

I was asked to operate a new type of magnetometer, the first electronic (ring core) magnetometer made in Japan. The purpose was to observe magnetic changes of the equatorial electrojet (the strong electric current along the magnetic equator), which was expected to be reduced by the eclipse (because of reduced conductivity of the ionosphere by the reduced solar X-rays).

It so happened that just one day before the eclipse, the magnetometer broke down. On the uninhabited atoll, there was no tool to repair it (nor any electricity). Thus, I brought a large battery set from the ship, and by short-circuiting it (with big sparks), I could fix the magnetometer before the eclipse time. I thought that this was like an appendix operation by myself in an uninhabited island.

The magnetometer recorded the expected change on a recording paper before my eyes (not in a dark room); fortunately, it was a very quiet (magnetically) day, so that the eclipse effect was only change superposed on the slow daily change due to the equatorial electrojet.

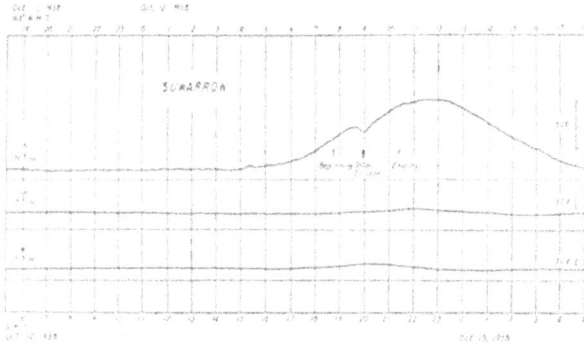

Figure 2.5 Magnetic record of October 12, 1958 at an atoll in the South Pacific Ocean. It was fortunately a very quiet day, so that the eclipse effect (a small dip) of the equatorial electrojet was clearly recorded.

There, I witnessed the solar corona for the first time. It was mysterious silky light. I was also asked to sketch the corona. Since the totality lasted 7 minutes, I had enough time to photograph the eclipse, too. At that time, I did not know anything about the solar corona, but my interest in it has remained; the solar corona is the subject of Section 7.2. It was an unforgettable adventure, three months across the Pacific Ocean on a 600-tonnage ship.

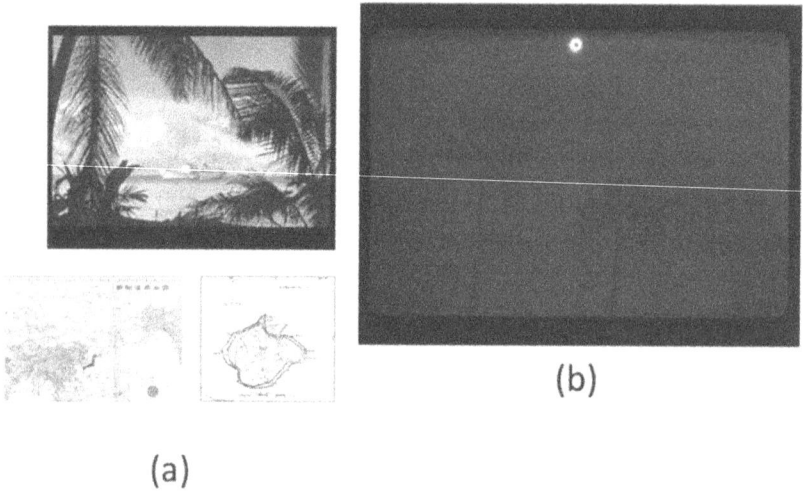

Figure 2.6 (a) The "Oshoro-maru" in the lagoon. Solar eclipse observation at a coral reef in the South Pacific Ocean (red dot on the map). (b) The total eclipse photo I took; in order to put palm trees as the background of the image, I had to go into a shark-infested water.

Soon after returning from the South Pacific Ocean, I arrived in Fairbanks, Alaska on December 13, 1958. I met Chapman at the Geophysical Institute (GI) of the University of Alaska in February 1959.

2.3 Main phase of geomagnetic storms

(a) Chapman's suggestion

Chapman's theory of the formation of the magnetosphere was successful in explaining what happens when solar gas collides with the earth's dipole and how the solar gas (plasma) flow forms a comet-shaped magnetosphere (Section 1.4c). This collision causes the storm sudden commencement (ssc): Chapman and Ferraro (1931). As a next step, he had been trying to understand the cause of a large decrease of the earth's magnetic field, the *main phase* of geomagnetic storms (Figure 2.7). This was the problem he worked since 1931 (so he told me).

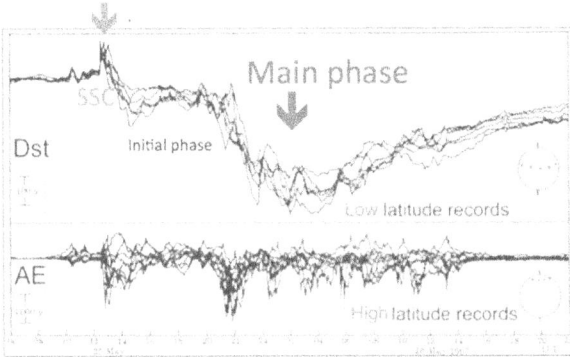

Figure 2.7 The standard development of geomagnetic storms. A typical geomagnetic storm (May 25-26, 1967). In the upper part, magnetic records from many low latitude stations are superposed; its average is the Dst index. In the lower part, records from several high latitude stations are superposed; the distance between upmost trace and lower most trace is the AE index. Note the difference of the scale in the upper and lower parts; γ = nT. In the early days, we had to collect individual magnetograms from many stations, digitize and superpose them to make the Dst and AE indices by ourselves.

The intensity of the plasma flow is given in terms of the kinetic pressure (= $(1/2)\, mnV^2$) of the solar wind. It is proportional to the magnitude of the sudden storm commencement (ssc), being typically 25 nT.

It is now understood that the sudden storm commencement indicates the impact of arriving the *shock wave* advancing ahead of the solar gas, now called "coronal mass ejection (CME)" or "magnetic cloud (MC)", because the solar wind and the magnetosphere are permanent. CMEs are described in Section 7.1 (h).

The impact of the shock wave results in the compression of the magnetosphere mainly at its front side and a world-wide simultaneous increase of the magnetic field (ssc); Figure 2.8.

41

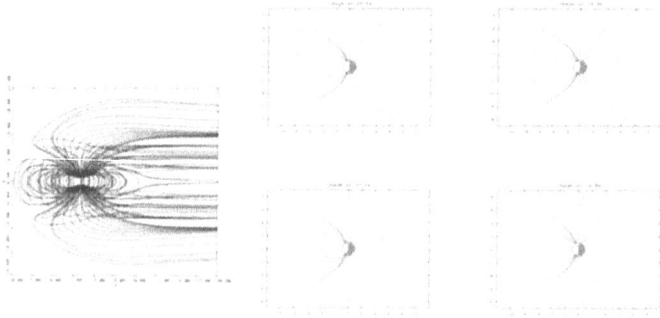

Figure 2.8 The compression of the magnetosphere by an intensified solar wind. When the shock wave associated with CME/MC impacts on the magnetosphere, it is "compressed", causing the sudden storm commencement (ssc), the Chapman-Ferraro theory (Section 7.1 [h]).

After succeeding to explain the storm sudden commencement (ssc), the next question by Chapman was how solar particles can get into the magnetosphere, causing a large decrease of the earth's magnetic field, which is called the *main phase* or the Dst decrease; Figure 2.9.

Although Chapman and Ferraro (1933) could not succeed in explain how the solar plasma can get into the magnetosphere, they suggested that the main phase is caused by a ring of westward flowing electric current within the magnetosphere and named it *"ring current"*.

Thus, the entry of solar particle into the magnetosphere had become a major question at that time, which Chapman and many others tried to find; Figure 2.9. However, they were not successful. Thus, Chapman asked me to succeed his study.

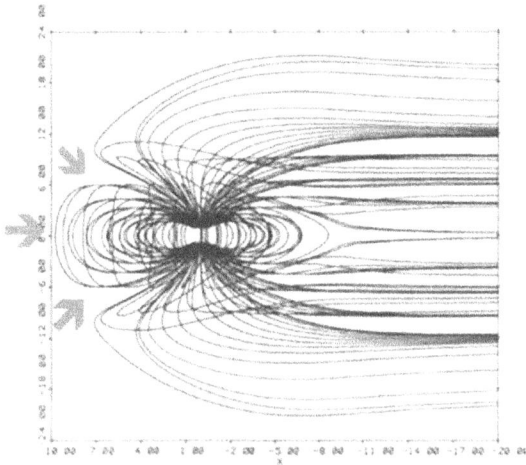

Figure 2.9 Chapman and many other theorists tried to find a way for solar wind particles to get into the magnetosphere to produce the ring current and the main phase, but could not find the entry process.

Thus, when I met him first, his suggestion was to study how solar wind particles can get into the magnetosphere. In a more general term, he was wondering what processes inside the comet-shaped magnetosphere could produce the main phase of geomagnetic storms and the aurora. In other words, how the magnetosphere can convert the solar energy carried by the solar gas flow into geomagnetic storms and auroral phenomena within the magnetosphere, so he told me. Actually, space physicists are now working specifically on this problem, as explained in Chapter 6.

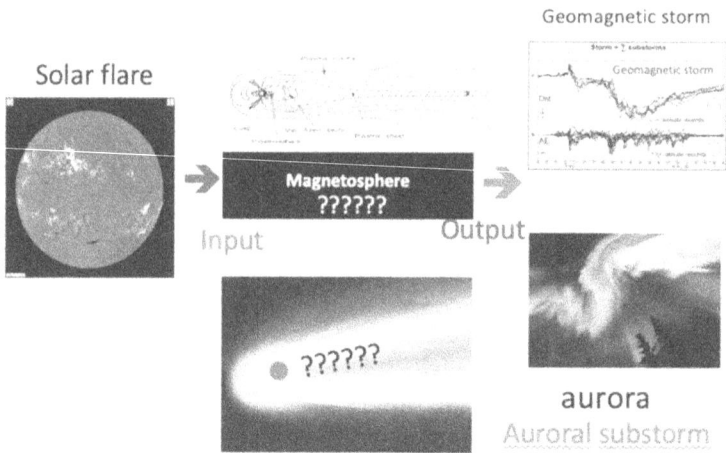

Figure 2.10 Chapman succeeded in explaining the formation of the magnetosphere. He suggested that our task was to study what processes in the magnetosphere produce geomagnetic storms and the aurora.

(b) The "Unknown factor"

When Chapman suggested this problem to me, I told him that I did not know enough about geomagnetic storms, so that I wanted to learn about them first by examining large number of records before studying the suggested problem. Since I worked at the magnetic observatory of my old university as a part-time job (Section 2.1), I was familiar with magnetograms and had no problem in handling them (but not knowing what they are indicating). I collected magnetograms of many geomagnetic storms from many stations in the world. In those days, they sent me the originals or their copies, so that I had to dig itize them. An example of magnetic records from low and middle latitude station is shown in Figure 2.11.

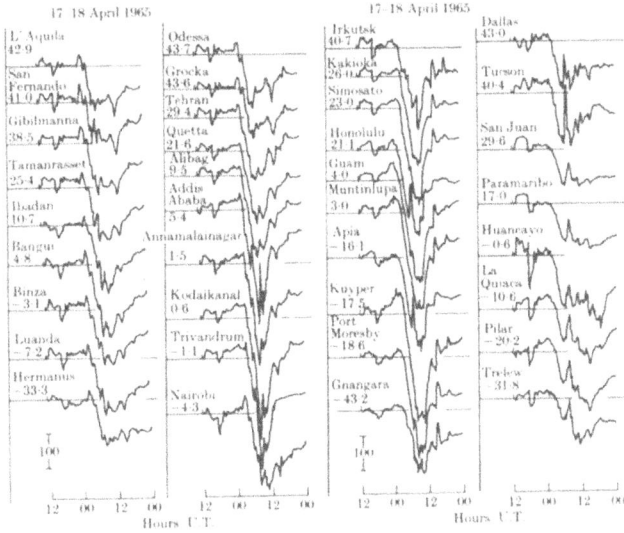

Figure 2.11 Magnetic variations during a geomagnetic storm of April 17-18,1965 at different longitude sectors (local time) and latitudes, from low latitude. Note that the main phase is greater in the evening sector (see. Figure 2.20). Unit: γ= nT.

After examining a large number of geomagnetic storm records, I found that geomagnetic storms develop in a great variety of ways, not necessarily in the standard way as considered earlier (Figure 2.7). In some cases, after a large sudden storm commencement (indicating the arrival of intense solar gas clouds), no main phase developed; such disturbances were thus not included in the list of geomagnetic storms (they are called a *sudden impulse*, *si*).

Since the magnitude of ssc indicates the intensity of solar gas flow, I found that the magnitude of the main phase is not related to the intensity of the solar gas flow. This is not in agreement with the theory by Chapman and Ferraro (1931).

In other extreme cases, an intense main phase occurred without sudden storm commencement (without the impact of a strong solar wind; they are called *gradually commencing storms*, 'sg'); such disturbances were also not listed as geomagnetic storms. There are a great variety of cases between the above two cases.

45

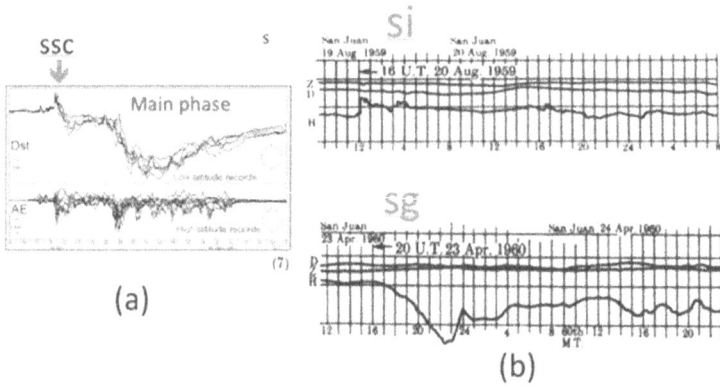

Figure 2.12 (a) The standard (or 'typical') geomagnetic storm. (b) Typical *si* without the main phase and *sg* disturbances without ssc.

I was very confused by the great variety of development of geomagnetic storms. Is there any way to put them in order? Eventually, I found one way. *It is not to consider the storm commencement as the onset of geomagnetic storms as far as the main phase is concerned.* This was a completely *unthinkable* idea at that time, because it had been so firmly believed that geomagnetic storms were caused solely by the *impact* of solar plasma and that an intense solar wind (recorded by the magnitude of sudden storm commencement) causes an intense geomagnetic storm.

Finally, I classified them by putting the *two **simplest** cases*, *si* at the top and *sg* at the bottom. This classification suggested me that something other than a stream of protons and electrons (plasma) in the solar wind is causing the main phase. This was also absolutely *unthinkable* at that time.

46

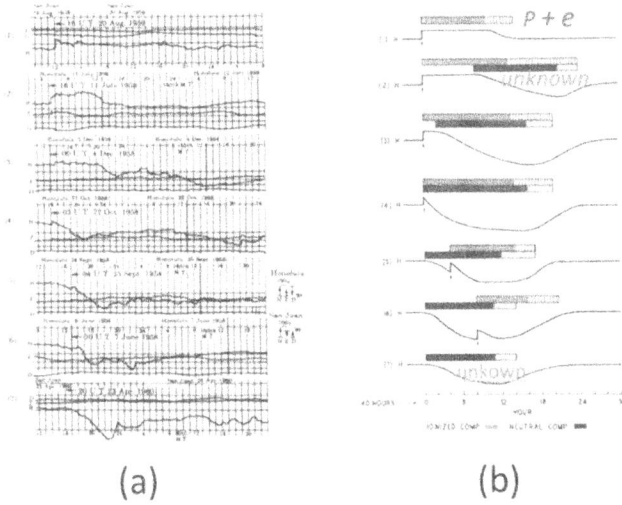

Figure 2.13 (a) My classification of the variety of geomagnetic storms. (b) My interpretation of (a). The top one is caused by a plasma consisting of protons and electrons (*p* + *e*). The bottom is caused by "unknown" factor. In other storms, "unknown" factor arrived independent of ssc.

Another interesting example I found is a long-lasting initial phase (7 hours), which is then followed by the *simultaneous* onset of the main phase and auroral activities. Such an example suggests also the arrival of the something other than protons and electrons about 7 hours after the arrival of solar plasma.

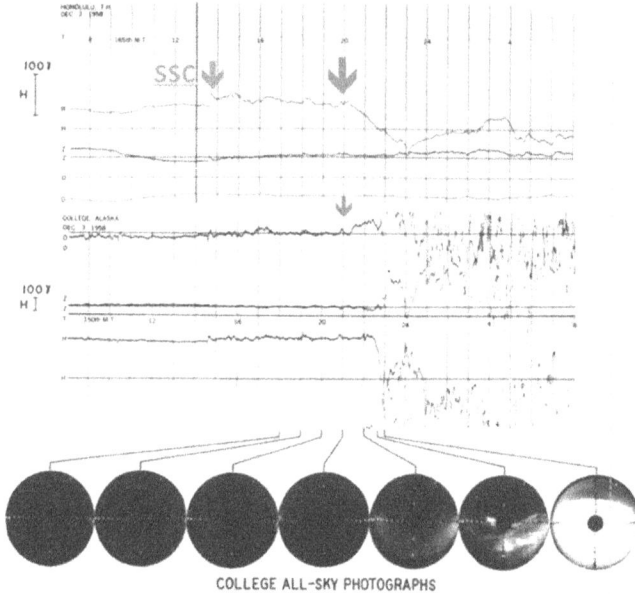

COLLEGE ALL-SKY PHOTOGRAPHS

Figure 2.14 An intense geomagnetic storm, which developed a large main phase and auroral activities together (all-sky camera images [GI]) only 7 hours after the storm sudden commencement. γ = nT. This suggested that the "unknown" factor arrived 7 hours after the ssc.

Based on these studies, I concluded that *there must be a "unknown" factor in the solar wind, which determines the development of the main phase.* At the time, everyone simply thought that the solar wind consists only of proton and electrons. Thus, the presence of "unknown" factor was also unthinkable idea.

After examining many more magnetic records with Chapman, he was also firmly convinced of the presence of the "unknown" factor in the solar wind which is responsible for the main phase; we took a walk together every possible day while discussing my result. I published my result by stating: "The variety of development of the storms seem to suggest some intrinsic differences between the solar streams far beyond what we would expect from a mere difference between their pressure" (Akasofu, 1963, p.125).

The three "unthinkable" ideas I encountered were:

(i)The sudden commencement (ssc) is not necessary the beginning of

48

geomagnetic storms.

(ii)The solar wind is not just a plasma of protons and electrons.

(iii)The solar wind contains something "unknown".

Obviously, there occurred serious objections, criticisms and controversies regarding our conclusion.

However, Chapman was willing to consider my three "unthinkable" results and accepted them in spite of his own idea of the standard geomagnetic storm (together with the fact that he was the first to treat the solar gas as a proton/electron plasma). He understood that his theory needed revisions in explaining the main phase and then strongly supported my results after he was convinced of my results.

Figure 2.15 With Sydney Chapman in1961 (GI). We published 20 joint papers.

Chapman presented our result during the *First International Conference on the Solar Wind* at the Jet Propulsion Laboratory (JPL) in 1964. After he came back, I asked him how our results were received. He told me that there were some comments. In fact, immediately after his talk, Jim W. Dungey suggested that the "unknown" factor might be the southward component of the interplanetary magnetic field (IMF [-Bz]) by stating: "I think the main phase of the storm is caused by something in the high-pressure gas, and the best guess is a strong southward interplanetary field. However, this is a question still to be answered." (From "*The Solar wind*,": Mackin and Neugebauer, 1966,

p. 272).

Soon afterward, a satellite observation by Don H. Fairfield and Larry J. Cahill showed that this was indeed the case (Fairfield and Cahill, 1966).

I recall also that Jim Dungey and I had once a pleasant discussion on auroral substorms (Chapter 4) at a park in London. He was very much interested in my analysis of auroral substorms (Akasofu, 1964a); he used my figure (Figure 4.4) in his paper. In fact, he was the first theorist who showed interest in my auroral work. His theory was that the magnetic field lines are stretched by the solar wind back to the magnetotail and then return back, and this process could explain substorms; his idea was much discussed later by many people. I explained that my observation suggested an internal cause of substorms (much closer to the earth, not the magnetotail); my morphological theory is described in Chapter 6. In the end, he said: "We had a friendly disagreement." It was one of my memorable occasions with him.

Don H. Fairfield told me that Jim suggested that he read my 1964 paper and suggested Don to find my "unknown" factor, namely the IMF (-Bz); Section 2.3. Don's first paper on this finding was published as a report from the Goddard Space Flight Center (GSFC).

With regard to Dungey's IMF (-Bz), Gene Parker (1958) showed earlier that the solar wind drags out the solar magnetic field (Section 1.4e).

Later, Hakamada and Akasofu (1982) simulated the magnetic field lines of the IMF on the equatorial plane up to 2 au (Section 7.3d); the simulation showed two structures (called the "co-rotating structures"), which are the shock waves caused by non-uniformity of the solar wind speed and which cause weak geomagnetic storms (Section 7.3d). This pattern rotates once in 27 days with the sun (seen from the earth), causing two 27-day recurrent geomagnetic storms (as Maunder suggested in 1905; Section 1.4).

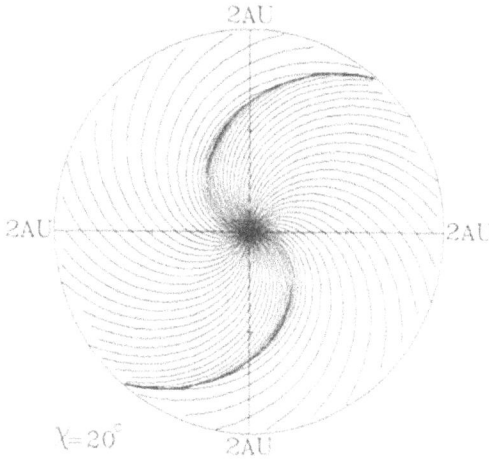

Figure 2.16 The solar wind drags out the solar magnetic field lines in a spiral form because the sun rotates (the equatorial plane). The figure shows the spiral field up to 2 au (1 au= the distance between the sun and the earth). Note the two co-rotating structures produced in the solar wind are caused by non-uniformity of the wind speed (Section 7.1 [vii]).

Along these stretched field lines of the solar magnetic field, various waves propagate (including the Alfven wave), so that these IMF field lines fluctuate, in the north-south and east-west directions (like swinging a rope fixed on the sun). What Dungey suggested was that when IMF field lines fluctuate southward, the main phase occurs. After Dungey's talk, Ian Axford commented that his theory (viscous interaction) could easily explain our result, although he withdrew it soon afterward.

However, it was not obvious at all for us and others at that time how the southward-oriented interplanetary magnetic field (IMF [-Bz]) is related to the cause of the main phase or how the IMF[-Bz]) could allow the solar wind energy to enter the magnetosphere. As is described in Section 5.2, we found that the IMF (-Bz) is the most crucial parameter in the auroral dynamo, which supplies the power for the aurora and geomagnetic storms. The degree of the function of the auroral dynamo depend on IMF (-Bz).

The nature of IMF [-Bz] is related the magnetic configuration of solar gas, coronal mass ejection (CME) or magnetic cloud (MC). The magnetic configuration of CMEs is one of the major subjects in Space Weather. In

51

Section 7.1 (h), Saito et al. (2007) got some idea on the magnetic field configuration by introducing electric current along magnetic field lines; see Figure 7.20.

(c) Ring current

Chapman and Ferraro (1933) assumed that solar particles somehow get into the magnetosphere, and they suggested that after the penetration, these particles would form a westward electric current. They named it the "**ring current**".

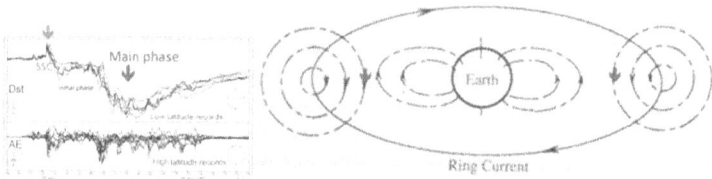

Figure 2.17 Left: A typical geomagnetic storm, showing the main phase (blue arrow). Right: Schematic illustration of the westward ring of current to explain the main phase (the Dst decrease).

(i)Magnetic field of the ring current explained the relationship between the main phase decrease (Dst) and the energy of the ring current on the basis of MHD (Dessler and Parker, 1959). However, in order to understand the nature of the ring current, we have to know how the charged particles form the electric current in the ring current.

When Van Allen of the University of Iowa reported the discovery of the Van Allen belts in 1959, Chapman thought that the belts could be their (Chapman-Ferraro) ring current belt. Chapman took me to visit Van Allen at the University of Iowa; they asked me to calculate the electric current in the belts and sent me to the Goddard Space Flight Center. There was an IBM 7090, a supercomputer at that time, and I was allowed to use it from the midnight to 4 am.

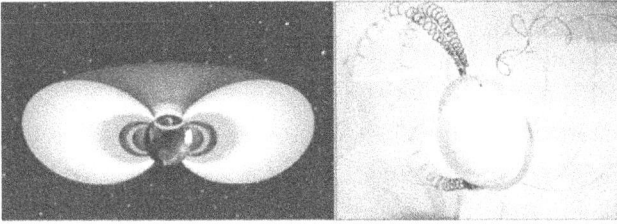

Figure 2.18 Left: The Van Allen radiation belt with the aurora. Right: Motions of protons (red) and electrons in the earth's magnetic field based on Stormer's study (Section2.2).

The study of Stormer (Figure 1.25) and the guiding center concept by Alfven (1950) showed how charged particles trapped in a dipole field move around. Based on my calculation, it was confirmed that high-energy particles trapped in the magnetosphere form the westward ring current, but the total energy in the Van Allen radiation belts was too small and cannot contribute much to the main phase decrease on the earth.

Thus, it was considered at that time that during geomagnetic storms, a new *storm-time Van Allen belt* must be formed (Akasofu and Chapman, 1961; Akasofu, Chain and Chapman, 1961). Such a storm-time proton belt was discovered by Lou Frank (1970) of the University of Iowa.

It was found that the current in the ring current is caused by circular guiding center motion (Alfven, 1950) and the gradient of distribution of ring current particles (diamagnetism), rather than the westward drift motion of protons; the ring current consists of two parts, the weaker eastward current in the inner side and stronger westward current in the outer side (Figure 2.19c), producing the magnetic field shown in Figure 2.19 (b); (Akasofu, Cain and Chapman, 1961). Larry Cahill (1966) confirmed our simulated magnetic field between the earth and 10 Re by a satellite. Thus, the presence of the storm-time ring current as the cause of the main phase decrease was confirmed. The numerical study of the ring current became the main contents of my Ph.D. thesis.

However, the question of how the ring current is formed was another major subject at that time. We had no idea from where the protons come from. As we describe in this section (iii) and Section 6.7, the ring current particles are mainly O^+ ions from the ionosphere during magnetic storms (although protons are also present), which are ejected from the ionosphere during auroral substorms (Section 6.8).

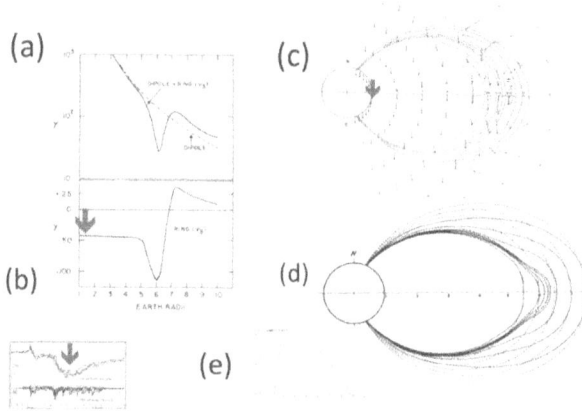

Figure 2.19 (a) The distorted dipole field by the ring current as a function of distance from the earth. (b) The ring current field. (c) The magnetic field vectors produced by the ring current. (d) Distorted (inflated) dipole field by the ring current. (e) The assumed distribution of protons in the belt [which was confirmed later by a satellite] (Akasofu, Cain and Chapman, 1961). The blue arrows indicate the main phase decrease.

(ii) Initial asymmetry of the ring current

During the initial study of the main phase (Figure 2.11), I found that the ring current is very asymmetric around the earth (in longitude) at an early epoch of the main phase; the main phase decrease is much larger (twice as much) in the evening sector; Figure 2.20.

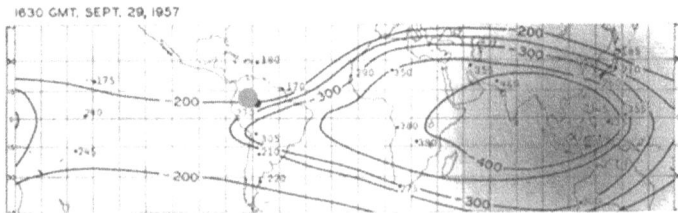

Figure 2.20 Asymmetric development of the main phase. The main phase decrease is asymmetric, largest in the evening sector by a factor of 2. The red dot indicates the subsolar point (Akasofu and Chapman, 1964).

Chapman and I speculated that protons enter the ring current *from the night side* and drift westward, making initially the main phase larger in the evening sector (Akasofu and Chapman, 1964).

This inference was found to be correct based on the satellite observation by DeForest and McIlwain (1971). They showed that protons (later found to be mainly O^+) enter the ring current belt from the midnight sector and shift westward. Thus, we began to understand a little more on the cause process of the main phase. The formation of the ring current had to wait until our substorm process advances: Section 6.7.

(iii) Two kind of ring current particles

We also found that the recovery phase cannot be expressed by one exponential function and thus speculated that there are two kinds of particles in the ring current belt (Akasofu, Chapman and Venkatesan, 1963). This inference was confirmed by the discovery of O^+ ions in the ring current belt by Shelley and Johnson (1972).

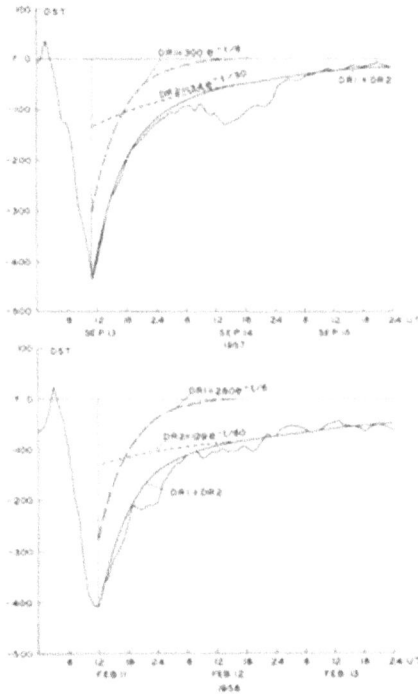

Figure 2.21 The two magnetic records indicate that the recovery phase of geomagnetic storms cannot be express by one exponential function, suggesting the presence of two kinds of ring current particles of different decay rates (Akasofu et al.,1963).

The discovery of O^+ by Shelley and Johnson in 1972 was one of the unexpected and greatest discoveries in magnetospheric physics, because everyone had thought until that time that ring current particles were all solar wind protons. They found that the ring current particles during *intense* geomagnetic storms are ionospheric oxygen ions (O^+), namely particles *originated in the ionosphere*, not in the sun; there is no O^+ in the solar wind.

Thus, the clue of the formation of the ring current must be related to the question of how energetic ionospheric ions (O^+) can be produced and injected into the ring current belt. In order to answer this question, we had to wait for the development of a study of auroral substorms (Chapters 5 and 6, specifically Section 6.8).

Later, it became possible to "see" the ring current when O^+ ions interact with other particles by charge exchange (Fok, 2003).

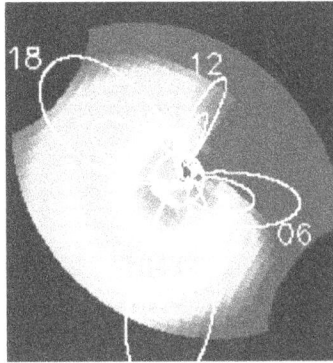

Figure 2.22 The ring current observed by optical emission produced by a charge exchange of ring current particles (Fok, 2003).

(d) Main phase decrease and polar magnetic disturbances

When I was examining low latitude magnetograms (for the main phase study), I had an opportunity to examine simultaneous College (Fairbanks) magnetograms and was surprised by the fact that an intense main phase is associated with a frequent occurrence of intense polar magnetic disturbances. Until the early 1960s, geomagnetic storms and polar geomagnetic disturbances (including auroral activity) were thought to be independent phenomena and thus almost independently studied. I checked this fact with Chapman.

Chapman and I concluded that intense geomagnetic main phases are associated with frequent occurrence of intense polar magnetic disturbances (representing auroral substorms), and weak geomagnetic storms (weak main phase) are associated with weak polar disturbances; Figure 2.23. In fact, *we concluded that intense main phase is the period when intense polar disturbances occur frequently.* This finding was published in Journal of Geophysical Research (Akasofu and Chapman, 1963a,b). However, it was not known at that time how the two facts are related. A possible relationship is discussed in Section 6.8.

This observation has led to our understanding that auroral substorms are responsible to the cause of the main phase and the formation of the ring current by producing O^+ ions. However, the understanding of this finding had to wait until the cause of auroral substorms and the double layer associated with it became more advanced (Section 6.8).

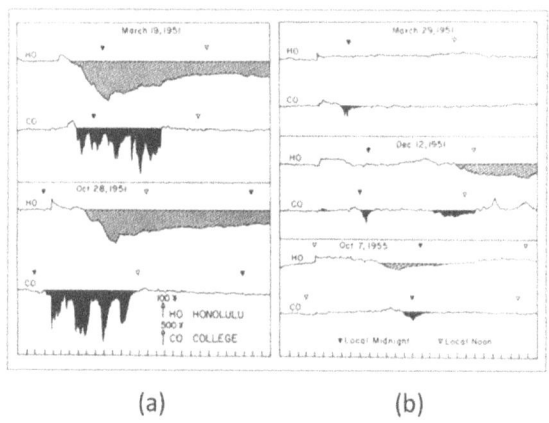

(a)   (b)

Figure 2.23 Comparison of geomagnetic storms of an intense main phase with intense polar magnetic disturbances and storms of weak main phase with weak polar magnetic disturbances; both Honolulu (low latitude) and College, Alaska (high latitude) records are shown together.

Episodes

(1) University of Iowa

The University of Iowa was one of the most active places in space physics in those days. It was truly remarkable that Van Allen and his group had led space physics of the United States by launching many satellites. However, apparently, the University of Iowa was not well known internationally in those days. Van Allen told me a story (twice) that he invited the director of the Jodrell Bank Radio Astronomy Observatory in England during a scientific conference. Imitating the director's tone, Van Allen told me his response was: "Where is Iowa?"

Figure 2.24 Left: With James Van Allen and his group at the front of the newly built Van Allen Hall (the University of Iowa). Right: Van Allen, kissing his radiation detector. Standing behind is Carl E. McIlwain (the University of Iowa).

Van Allen was awarded by Norwegian King for his discovery of the radiation belts. Many of us attended his awarding ceremony in Oslo and had a dinner with the King. Afterward, Van Allen told us: "Folks, it is for the longevity."

(2) Discovery of the mantle flow

During the search of the "unknown" parameter (Section 2.3b), I considered once neutral hydrogen atoms in the solar wind, because they can penetrate across the magnetosphere without any trouble and can become energetic ring current protons by charge exchange. Solar prominences seen by the $H\alpha$ line is emitted by neutral hydrogen atoms (Akasofu, 1964b).

When I had an opportunity to work on the solar wind at the Los Alamos Laboratory, I was examining the solar wind in their satellite records and found that the solar wind blows just *inside* the magnetopause (Akasofu et al., 1973). Since it had been considered that the solar wind blows only outside the magnetopause, I thought that the flow contained neutral hydrogen atoms. After a sleepless night, I found that the solar wind (plasma) can penetrate into the magnetosphere, but only near the boundary of the magnetosphere.

This flow is now called the *mantle flow*. I still believe that a very small amount of neutral hydrogen atoms might exist in the solar wind.

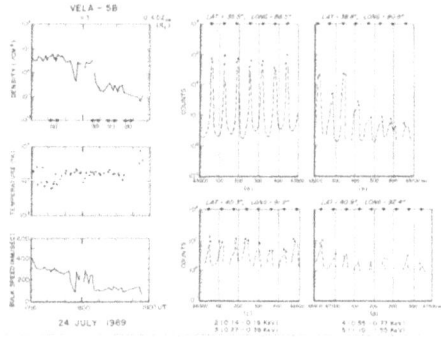

Figure 2.25 The solar wind flow just inside the magnetopause. As the detector spins, the solar wind outside the magnetopause shows a larger modulation, but a smaller modulation inside because of its weak flow (Akasofu et al., 1973).

(3) Early days at the Goddard Space Flight Center

When I was working at the GSFC for the calculation of the magnetic field of the ring current, a satellite launched by the GSFC crossed the magnetopause for the first time. At the GSFC, Jim P. Hepper, Norman F. Ness and Joe C. Cain were wondering what happened without realizing first that the satellite got out the magnetosphere, namely crossing the magnetopause. It was in such a way that the magnetosphere had been explored in those days. There were only two buildings at the GSFC at that time (even without gate).

Figure 2.26 Left: Working with the IBM 7090 computer at the Goddard Space Flight Center in 1960 (NASA). Right: Preparation of one of the early satellites for exploring the magnetosphere (NASA).

By extensive satellite observations in the 1960s, 1970s and 1980s, the magnetosphere had been explored, and magnetospheric physics was also well developed.

(4) Two AGU publications

American Geophysical Union published two publications collecting pioneering papers in space physics. I contributed to both.

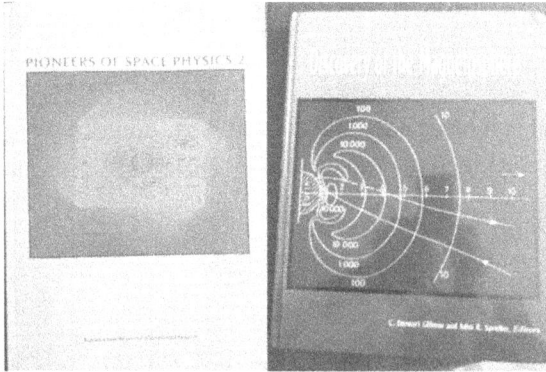

Figure 2.27 Two AGU publications assembling articles by early researchers in magnetospheric physics.

At that time, it was still not yet understood at how the internal structure of the magnetosphere is related to processes which can cause auroral activities. This is partly because the study of auroral activities was only at an early developmental stage.

(5) Remembering Sydney Chapman

Officials of the University of Alaska invited Chapman to join the Geophysical Institute. I believe that there was another reason why he was interested in the Geophysical Institute. In 1952, there was a conference in Canada (organized by the US Air Force Laboratory in Boston) on the aurora (the first international auroral conference after the WW II). There, Stormer criticized Chapman's theory by saying Chapman's theory cannot explain the aurora at all, but his theory of the electron beam could explain many detailed features of the aurora (from the transcript). Chapman presented "An auroral theory," but the content was only the Chapman-Ferraro theory of storm sudden commencement (ssc). In fact, Chapman told me that he made the mistake there by having the title "An auroral theory".

In addition, the Chapman-Ferraro theory (Section 1.4) had not necessarily been accepted as a theory of geomagnetic storms at that time. There was a new report of detecting the hydrogen emission in the aurora during the conference, and it was thought that it supported Chapman's theory rather than Stormer's theory of electrons. However, I believe that Stormer's criticism was one of the reasons for Chapman to come to Alaska, but I missed the chance to ask him about it.

One thing Chapman repeatedly told me was what Julius Bartels, the co-author of *Geomagnetism*, told him: "A person who criticized your paper at least read your paper."

Figure 2.28 Sydney Chapman and Carl Stormer were shaking hand during a conference in Canada (Chapman).

Sydney Chapman was born in Eccles, a suburb of Manchester, England in 1888. When Chapman was 14, his father took him to a builder's merchant. He told Chapman about a plumber who made a fortune in America; Chapman once jokingly mentioned that he might have also been rich if he'd emigrated to America and became a plumber.

Then, his father took him to an engineer, who suggested he attends a technical school for two years and even go to Manchester University. He was chosen as 15th of 15 students there. He wondered many times in his later years what might have happened if he were not chosen (so told me a few times); I told him that the beginning of magnetospheric physics would have been delayed for many years.

When Chapman was 22 years old and a student at Cambridge student, F. W. Dyson, director of the Greenwich Observatory, visited him, introducing himself as Astronomer Royal, and offered Chapman a job as an assistant. It was a very prestigious and promising job, and thus Chapman reported it proudly to his father; his first job was to improve their magnetic observatory.

It seems that his scientific life at the Greenwich Observatory was crucial in many ways, because his life-long works in various fields (including geomagnetic storms and the lunar atmospheric tide, which he discovered) began at the Greenwich, although he went back and forth between the Observatory and the Trinity College in Cambridge, the Manchester University, the Imperial University, and finally the Oxford University (1946-1953); Chapman had a great affection for the Trinity College and took me there.

Chapman told me that their *"Geomagnetism"* book (Chapman and Bartels 1940) was getting out of date, so that he and I should publish a new book titled *"Solar-Terrestrial Relationship."* It was the time when the field of solar-terrestrial relationship and space physics were rapidly progressing, so that it took about one month to revise one chapter and thus took one year to revise the whole manuscript. Repeating the revisions a few times, it took several years to complete it.

Unfortunately, Chapman passed away in 1970 before its completion in 1972. It was finally published from the Oxford University Press. I dedicated it to Katharine Chapman, because of her great support of her husband. She was also encouraging me and my wife Emiko in many ways during my early days at the Geophysical Institute.

Walter O. Roberts, president of the National Center of Atmospheric Research (NCAR), suggested that we publish a book for Chapman's 80th birthday by collecting stories about Chapman from as many people as possible. I had to ask Katherine for her help in finding potential contributors; unfortunately, but also fortunately, one day, she left the address of people for possible contributors on her desk, and he found our plan and helped us.

We collected about 600 signatures and about 100 contributing remarks and episodes, including one of his high school friends and even an old teacher. The book was published with the title "Chapman, Eighty *from his friends*" by the University of Colorado Press (Akasofu, Fogle and Haurwitz, 1968).

The book, co-authored by Bernhard Haurwitz and Benson Fogle with me, contains Chapman's three talks (transcribed) about his life: the first one on October 8, 1965 at NCAR: the second on December 16 at the NCAR in 1966; the third June 30, 1967 at the Geophysical Institute of the University of Alaska Fairbanks. His talks are mainly his life story; he talked about many people he was acquainted with, including Rutherford, Lamb, Reynolds, Lord Rayleigh, Hale, Eddington, Massey, Bertrand Russell and many others.

The most frequently expressed words by his friends, regardless of age and scientific status, are a great admiration of his personality: kindness, warmth, sympathetic, polite, honesty, graciousness, modesty, humbleness, competent, inspirational, encouraging, man of integrity, incorruptible, simplicity and courageous.

One of the best stories conveying his personality in the book was provided by Subrahmanyan Chandrasekhar, a famous astrophysicist. When Chapman was honored by the Royal Society of London, his friend offered to drive him back home after the ceremony, since the evening was dark and foggy. Chapman (formally and fully dressed) declined, saying "I came by bicycle."

I remember that he told me at one occasion: "Syun. I want to let you know I declined Knighthood." His reply to my question why was "Katharine does not want to be called 'Lady Chapman'."

Chapman asked many people to call him Sydney. Being from Japan, I could not do it, as was the custom there. I called him "Dr. Chapman."

Because he asked it so often, I finally told him that when my age reaches half of his, then I would call him by his first name; he smiled, and did not press further. He died at 81 in 1970, when I was 40. I regret I lost that opportunity to call him "Sydney" forever.

I had a great difficulty in judging Chapman's place in the history of science. I did not know that Arthur Eddington (a great astronomer and author of the classic treatise "*The Internal Constitution of Stars*") worked side by side with him at the Greenwich Royal Observatory.

Chapman used to swim 25 laps every day possible at the pool of the University of Alaska. When I swam with him, he wanted to be always ahead of me. In many international conferences, the responsible organization reserved a good hotel for him, only to find in panic that he was not there; when they asked me where he might be, my response was always: "You will find him in a pool at the YMCA." He did not like to stay at an expensive hotel.

I published two papers titled "The scientific legacy of Sydney Chapman" (Akasofu, 2011) and "Sydney Chapman: A biographical sketch based on the book "Chapman, Eighty, from his friends" (Akasofu, 2020).

My motive

My study of geomagnetic storms began, when Chapman suggested a theoretical study of the main phase of geomagnetic storms. Instead of considering theoretically the main phase on the basis of the most authoritative book '*Geomagnetism*' by Chapman and Bartels (1940), I wanted to see *actual* magnetograms; I was familiar with *handling* magnetograms (although I did not understand what is recorded; Section 2.1).

I was very surprised by the fact that geomagnetic storms develop in various ways. This led me to consider that the solar gas (the solar wind) contains something other than protons and electrons, namely "unknown" factor (Section 2.3). Such an idea was absolutely "unthinkable" at that time.

After all, it has taken almost 50 years to partially solve Chapman's problem on the main phase of geomagnetic storms (Section 6.8).

References

Akasofu, S.-I., The development of the main phase of magnetic storms, *J. Geophys. Res., 68*, 125-129, 1963a.

Akasofu, S.-I., The main phase of magnetic storms and the ring current, *Space Sci. Rev., 2,* 91-135, 1963b.

Akasofu, S.-I., 1964a, The development of the auroral substorm, *Planet. Space Sci., 12,*273-282. https://doi.org/10.1016/0032-0633(64)90151-5

Akasofu, S.-I., 1964b, The neutral hydrogen flux in the solar plasma flow-1, Planet Space Sci., **12**, 905.

Akasofu, S.-I., Fogle, B. and Haurwitz, B., 1968, *Sydney Chapman Eighty,* Published by the National Center for Atmospheric Research and the Publishing Service of the University of Colorado.

Akasofu, S.-I., 2011, The Scientific legacy of Sydney Chapman, EOS, *92*, 281, doi: 10.1029/2011EO340001.

Akasofu, S.-I., 2020, Sydney Chapman: A biographical sketch based on the book "Chapman Eighty, from his friends", Perspectives of Earth and Space Scientists, 2, e2020CN000135. https//doi.org/10.1029CN000135.

Akasofu, S.-I., and Chapman, S., 1961, The ring current geomagnetic disturbance, and the Van Allen radiation belt, J. Geophys. Res., **66**, 1321.

Akasofu, S.-I., Cain, J. C., and Chapman, S., 1961, The magnetic field of a model radiation belt, numerically computed, J. Gephys. Res., **66**, 4013.

Akasofu, S.-I. and Chapman, S., 1963a, The development of the main phase of magnetic storms, J. Geophys. Res., **68**, 125.

Akasofu, S.-I. and Chapman, S.,1963b Magnetic storms: The simultaneous development of the main phase (DR) and of polar magnetic substorms (DP), J. Geophys. Res., **68**, 3,155.

Akasofu, S.-I., Chapman, S and Venkatesan, D.,1963, The main phase of great magnetic storms, J. Geophys. Res., **68**, 3,345.

Akasofu, S.-I. and Chapman, S., 1964, On the asymmetric development of magnetic storm fields in low and middle latitudes, Planet Space Sci., **12**, 607.

Akasofu, S.-I., Hones, E. W. Bame, S. J., Asbridge J. R. and Lui, A. T. Y., 1973, Magnetotail boundary layer plasmas as a geocentric distance of 18 Re, Vela 5 and 6 observations, J. Geophys. Res., **78**, 7257.

Alfven, H., 1950, *Cosmical Electrodynamics*, Oxford Univ. Press.

Cahill, L. J., 1966, Inflation of the inner magnetosphere during a magnetic storm, J. Geophys. Res., **71**, 4505.

Chapman, S. and Ferraro, V. C. A., 1931, A new theory of magnetic storms, Terr. Mag. Atmos. Elect., **40**(4) 349.

Chapman, S. and Ferraro, V. C. A., 1933, A new theory of magnetic storms, Ter. Mag. Atmos. Elect. **38**(2), 79. https://doi.org/10.1029/TE38002p00079.

Deforest, S. E. and McIlwain, C. E., 1971, Plasma clouds in the magnetosphere, J. Geophys. Res.,**76**, 3587.

Dessler, A. J. and Parker, E. N.,1959, Hydromagnetic theory of geomagnetic storms, J. Geopys. Res., **64**, 2239.

Dungey, J. W., 1966, Solar-wind interaction with the magnetosphere. Particle aspects, 243-255, in The Solar wind, by R. J. Mackin, jr. And M. Neugebauer, Jet Propulsion Laboratory, Pasadena, California.

Frank, L. A., 1970, Direct observation of asymmetric increase of exstra terresstrial 'ring current 'proton intensities in the outer radiation zone, J. Geophys. Res., **75**, 1268.

Fok, M. C. et al., 2006, Impulsive enhancement of oxygen ions during substorms, J. Geophys. Res., **111**, A10222 doi:1029/2006JA011839

Fairfield, D. H. and Cahill, L. J., 1966, Transition region magnetic field and polar magnetic disturbances, J. Geophys. Res. **71**(1), 155, https://doi.org/10.1029/JZ071i00155

Gibert, William, 1540-1603, translated by P. Fleury Mottelay, Dover publication (paperback, 1958).

Mackin R. J. and Neugebauer, M.,1966, *Solar Wind*, Jet Propulsion Laboratory, Pasadena,

Shelley, E. G. and Johnson, R. G., 1972, Satellite observations of energetic heavy ions during geomagnetic storm, J. Geophys. Res.,**77**, 6104.

Van Allen, J. A. and Frank, L. A.,1959, Radiation around the earth to a distance of 107,400 km, Nature, **183**(4659), 430-434, https://doi.org/10.1038/183430a0

Chapter 3 Auroral oval: Distribution of the aurora

The location of the auroral belts, both in the northern and southern hemispheres, is fundamental in understanding a number of auroral and magnetospheric phenomena and processes. The auroral oval is the natural coordinate system. Many space physics phenomena are different, depending on inside, within and outside the auroral oval. In this chapter, we learn how it was determined and how important the auroral oval is in studying auroral phenomena.

3.1 Auroral zone

Elias Loomis established the auroral zone in 1860 by collecting all available information. The auroral zone had been firmly believed even during the International Geophysical Year of 1957-1958.

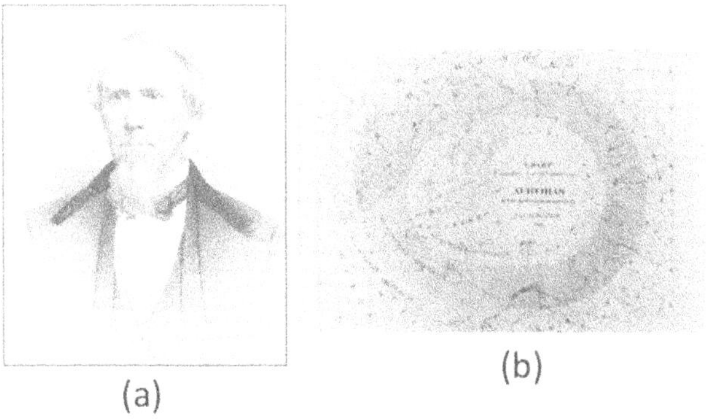

Figure 3.1 (a) Elias Loomis. (b) The auroral zone determined by Loomis.

Soon after I arrived in Fairbanks, I began to observe the aurora every night it was possible. Since Fairbanks is located in the auroral zone, I expected that I could observe the aurora as soon as the darkness set in (5 pm in midwinter). However, I found that I had to wait until about 9 pm to see it near the northern horizon.

Thus, the first problem I encountered on the aurora was that my *visual* observation of the aurora did not seem to agree with the well-established auroral zone by Loomis. I was not even an auroral physicist yet at that time.

3.2 Auroral oval

The curtain-shaped aurora is called an "arc" for a historical reason that norther people watched the aurora as an arc or arch near the northern horizon (Section 1.2). When I began to observe the aurora for the first time in Fairbanks, Alaska (65° north) in 1959, I noticed that an auroral arc appears almost always in the northern sky in the evening and shifts toward Fairbanks as the evening progresses. It then appears to go back to the northern sky in the morning.

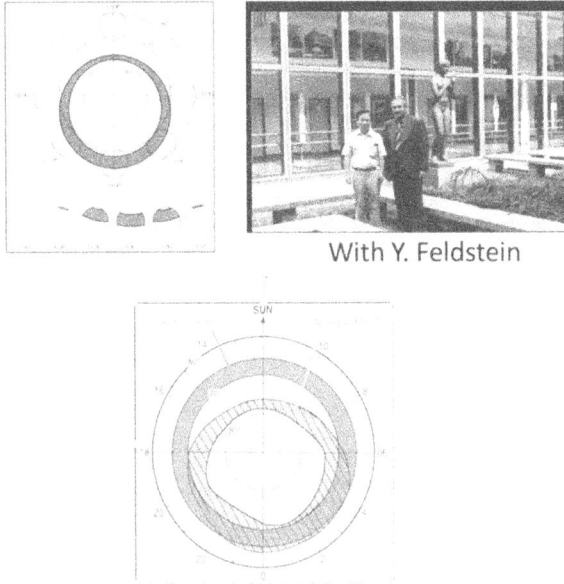

With Y. Feldstein

Figure 3.2 Upper left: View of the auroral oval at different local (magnetic) times. The circle indicates the field of view at Fairbanks, Alaska. Right: With Yasha Feldstein. Lower: The auroral zone (pink) and the auroral oval (green) determined by Y. Feldstein.

I went to the director's office (as a graduate student) and asked Chris Elvey, GI director at that time, about why the aurora did not appear overhead at any time of the night above Fairbanks.

Elvey told me that it had long been believed that the auroral arcs were formed at the center line of the auroral zone, 67° in geomagnetic latitude, and then shifted southward after they were formed. My colleagues said: "That is the way it is."

I then examined all-sky images at Fort Yukon (67°, the center line of the auroral zone) and found that the arc also appears first in the northern sky. Then, I found that even at Utqiagvik (the former Barrow, at 70°, the northern edge of the auroral zone), the aurora also appears first in the northern sky and then shifts in the southern sky in the late evening.

Although I was just a beginner of auroral studies, I thought something was quite wrong with the well-established auroral zone. But at that time, I was busy in my study of geomagnetic storms with Chapman and thus forgot about my observation, until I found a paper by Yasha Feldstein (1963). He showed that the belt of the aurora has an oval-shape (not an annular ring like the auroral zone in geomagnetic coordinate). He called the auroral belt the "*auroral* oval;" the day side of the auroral zone is displaced toward higher latitude by more than 10°, so that its midday part of the location is about 75° in latitude or higher (instead of 65° like the auroral zone).

Since his results clarified my question on the auroral *zone,* I wrote to Feldstein that I agreed with his results. Then, he wrote me back, mentioning his difficulty in convincing his colleagues about auroral oval. I myself had a great difficulty of convincing my colleagues about the oval shape. As my observation agreed with Feldstein's oval, I met him in Moscow in 1968 to discuss how to confirm the oval shape. This was the beginning of the close relationship with Yasha.

Incidentally, at the Geophysical Institute, we had an opportunity to examine the midday aurora (based on the concept of the auroral zone) when we had a chance to observe the total solar eclipse of July 20, 1963. Thus, in addition to the observation of the eclipse, an additional observation of the midday aurora was planned, but the aurora was not observed.

3.3 Proving the auroral oval

I thought that one way to prove Feldstein's finding (auroral oval) was to set up a chain of all-sky cameras along the magnetic meridian line between Alaska and the northwestern tip of Greenland (located near the geomagnetic pole). This chain of all-sky cameras can scan the whole polar sky once a day as the earth rotates, like an azimuth-scan radar at an airport; such a chain of all-sky cameras is perhaps the largest scanning 'device' in the world. This study was supported my first grant from the National Science Foundation (NSF).

It was a difficult task to find the right locations in the Canadian wilderness in order to secure an appropriate power supply for the all-sky cameras to operate.

This observation proved the accuracy of the auroral oval, as Feldstein established; in midday hours, the aurora is located at about 75° in latitude. However, unfortunately, I could not convince many colleagues with the meridian chain result, since the auroral zone had been believed so long.

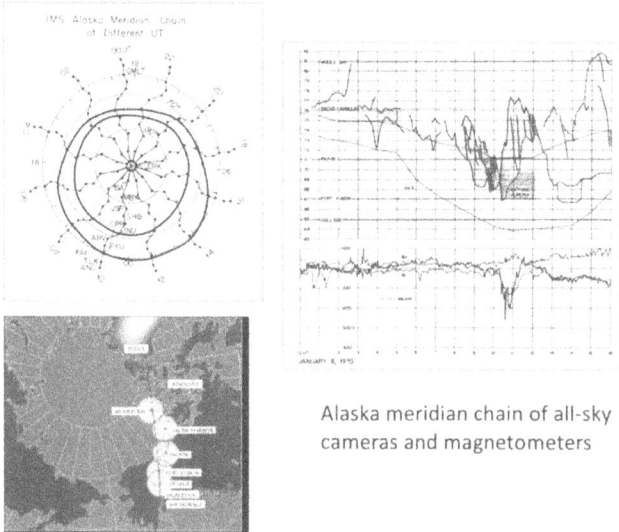

Alaska meridian chain of all-sky cameras and magnetometers

Figure 3.3 Left: Upper. Alaska meridian scanning all-sky network in geomagnetic longitude line; it scanned the whole polar sky by the chain of all-sky cameras in one day. Lower. The location of all-sky cameras; later, both Thule and Alert were added. Right: The location of the aurora in geomagnetic latitude-time coordinate, beginning from magnetic midday to early morning). The location changed from about 75° in the midday hours to about 65° in midnight hours.

At that time, the US Air Force was interested in auroral research for various strategic reasons. They sent me Col. A. Lee Snyder as a graduate student. He completed his Ph.D. thesis on the analysis of the Alaska meridian chain of all-sky cameras in 1972 (Snyder and Akasofu, 1972, 1974).

Further, an Air Force plane flew along the auroral arcs to confirm the shape of the oval; actually, they and I made a templet of the auroral oval on a map (with different degrees of auroral activities), which pilots could use in determining their position with respect to the auroral oval in their polar flights. In one of the test flights along the oval, I predicted their arrival time in their flight from

Boston to Iceland. I recall I was relieved when they called me that they arrived in Iceland at the expected time by following the templet; Buchau et al. (1970).

The accurate determination of the auroral oval is scientifically crucial, because it provides the *natural co-ordinate system*, along which many polar electromagnetic phenomena occur. Some phenomena occur only inside the oval (precipitation of energetic solar electrons and the resulting polar cap absorption). The auroral electrojet flows along the auroral oval, not along the auroral zone (Section 6.3[d]).

At that time, I had an opportunity of working at the physics department at the University of Iowa on satellite observations. There, I found that the auroral oval coincides with the intersection line of the outer boundary of the outer Van Allen belt with the ionosphere (Van Allen's study). I rushed into Van Allen's office to show him the result:

"You should know that the radiation belt particles are not auroral particles," he said.

"I know that," I said. "The location of the outer boundary of the radiation belt coincides with the auroral oval. That means that auroral particles stream into the ionosphere along the outer boundary of the outer radiation belt." This suggests that the origin of auroral particles is located at the boundary of the outer radiation belt in the equatorial plane, namely about 6 Re, not at a long distance in the magnetotail.

It was about that time when the magnetotail was discovered (Ness, 1965), and my first paper on auroral substorms was published (Akasofu, 1964). Thus, many theorists were interested in the relation between the aurora and magnetic reconnection in the magnetotail (Section 7.1 [vi]), paying no attention to the above finding (the intersection line of the outer boundary of the outer radiation belt on the ionosphere coincides with the auroral oval). It is crucial in determining the origin of auroral particles; it was unfortunate that such a simple, but crucial fact, was not considered by them. This fact is mentioned again in Section 6.2.

Van Allen immediately understood the importance of the auroral oval (the source of auroral electrons must lie along the outer belt of the radiation belt); the source region of auroral particles is crucial in understanding the aurora. He asked me to report the result in a publication from Department of Physics.

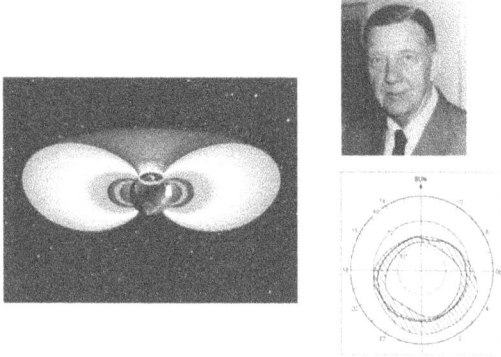

Figure 3.4 Left: The Van Allen radiation belt and the aurora. Right: James A. Van Allen (the University of Iowa). The lower figure shows that the auroral oval and the outer boundary of the outer radiation belt (line) determined by Van Allen. They the coincide well.

Fortunately, to my colleagues, this agreement between the auroral oval and Van Allen's study was a little more convincing the validity of the auroral oval than the results of the Alaska meridian chain operation of all-sky cameras.

One lesson I learned from this is that it is difficult to convince colleagues about a new auroral result by other auroral observations. It becomes more convincing with a completely different set of observations (for example, the outer boundary of the radiation belt in this case) obtained without considering (or without knowing) the relation with the aurora.

It so happened that my Physics Department report got the attention of Al J. Zumuda of Johns Hopkins University. Zumuda was determining the location of field-aligned electric currents (currents flowing along magnetic field lines) that are responsible for causing the aurora. His location of field-aligned currents in the polar region coincided with the auroral oval (Armstrong et al.,1973). This study confirmed further the validity of the auroral oval.

In 1978, Iijima and Potemura (1978) made a detailed study of the location of the field-aligned currents and confirmed that the field-aligned currents flow in and out from the auroral oval. We confirmed also their study on the basis of simultaneous all-sky images in Alaska and observed field-aligned currents by their satellite.

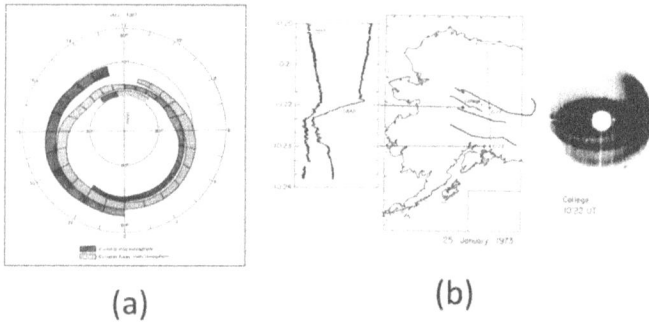

(a)   (b)

Figure 3.5 (a) The field-aligned current distribution determined by Iijjima and Potemura (1978). (b) The simultaneous observation of the aurora (black in the all-sky image) and magnetic field changes produced by the field-aligned currents). Satellite record showing field-aligned current, location of auroral arcs and the all-sky camera image.

Meanwhile, Cliff D. Anger, the University of Calgary, obtained the first auroral image from a Canadian satellite in 1972. It clearly showed the auroral oval shape. Cliff, A. Tony Y. Lui (my second post doc from Anger) and I published this observation together (Anger et al., 1973). This was the beginning of a long association with Tony until today.

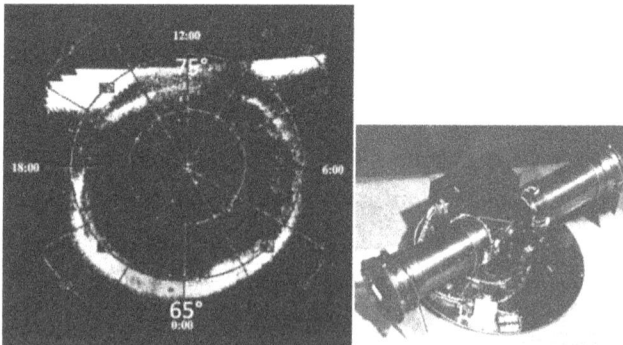

Figure 3.6 Left: First satellite image of the auroral oval, indicating that the dayside part is located at 75° (Anger et al., 1973). Right: The photometric instrument used by C D. Anger (C. D. Anger).

With the satellite image, the great controversies on the auroral oval soon faded. It took 9 years (from 1963 to 1972) to settle the controversy, although the above result can now be mentioned in one sentence: "The aurora appears along the auroral oval." in a textbook or monograph.

I learned how difficult it was to convince my colleagues with a new finding, and auroral specialists were the most skeptical on a new finding.

After all, the fact that an auroral arc appears first near the northern horizon and shifts southward every evening is due to the fact that the earth and an observer move under the fixed pattern of the auroral oval, not the aurora shifts toward us.

However, some of my colleagues were still suspicious about the oval shape and asked me to explain why the auroral oval does not have an annular ring shape.

I explained that the auroral oval appears along the belt surrounding the root of the linked field lines between interplanetary magnetic field (IMF) lines and magnetospheric magnetic field lines. This belt coincides with the auroral oval. The linked field lines are called the "open field lines" and the opened magnetosphere is called the "open magnetosphere" (Chapman's magnetosphere is called the "closed magnetosphere."). The area surrounded by the auroral oval is the "open region". Based on a simple magnetosphere model, it is possible to determine the distribution of the linkage point of the IMF field lines and magnetospheric field lines (Figure 3.7c).

It can be seen that the roots are surrounded by the auroral oval. This fact is well confirmed by the fact that energetic solar electrons (100 KeV) from intense solar flares enter in the region surrounded by the auroral oval by following the open field lines.

Thus, this simple model has become the basis of the auroral dynamo, which will be the foundation of Chapters 5 and 6.

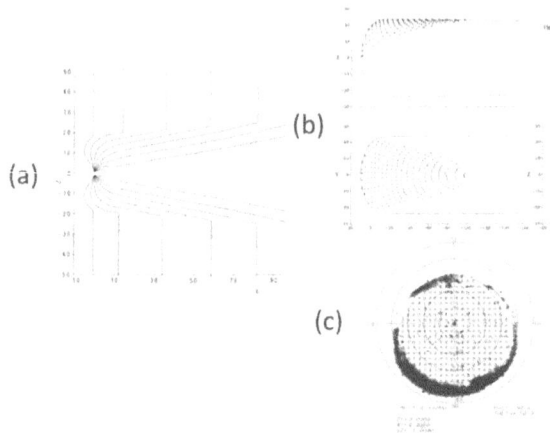

Figure 3.7 (a) Linked field lines between the interplanetary magnetic field and the magnetospheric field (Akasofu, Roederer, Corrick and Covey, 1981). (b) The linkage points of both field lines on the boundary of the magnetosphere; the side and upper views. (c) The "roots" of the linked field lines in the polar region. The auroral oval surrounds the roots.

The size of the auroral oval depends on the intensity of the IMF (-Bz) (or the total flux of the open field lines). The linkage points on the boundary of the magnetosphere becomes crucial in understanding the cause of auroral activity called the auroral substorm (Chapters 5 and 6).

When the IMF has the X or Y component, the roots of the open field lines in the polar regions differs from the IMF (-Bz) case. They are shown at the end of this chapter.

3.4 Equatorward expansion of the auroral oval

The auroral oval expands during intense geomagnetic storms. This is because IMF (-Bz) can be as large as -25 nT in coronal mass ejections (CMEs) [most of the time, -5 nT or less] and thus the open flux increases. As a result, the size of the oval increases.

However, this cannot explain why the aurora can be seen in the middle latitudes of the US as well as in London and Stockholm (Figure 3.8) during geomagnetic storms. As explained below, it is likely that the ring current changes the internal distribution of the geomagnetic field, when the main phase is large.

Chapman was very interested in the appearance of the aurora in lower latitudes. Akasofu and Chapman (1963) determined the relationship between the southern boundary of the auroral oval and the main phase decrease given by the geomagnetic index Dst (a measure of the main phase decrease (or the intensity of the ring current) in the US sector.

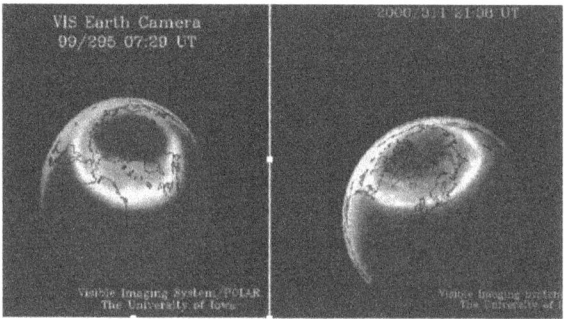

Figure 3.8 Enlarged oval during intense geomagnetic storms. Left: The US sector. Right: Europe sector (the University of Iowa).

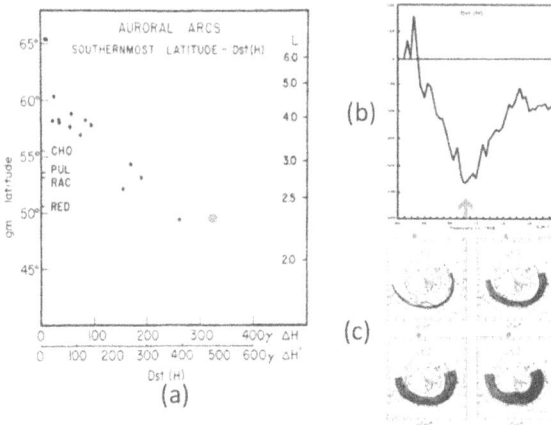

Figure 3.9 (a) The relationship between the latitude of the oval (the southern boundary) and the Dst index in the US sector; the red point indicates the location of the oval during the storm of February 11, 1958. (b) The Dst index of February 11, 1958 storm. (c) One of the most intense substorms during the same storm). A great auroral substorm occurred at

79

about the same time, covering the latitude from 50° to 68°.

A very large auroral oval during very *intense* geomagnetic storms is, however, impossible to explain by the intensity of IMF Bz (or of the open field lines) alone.

In fact, in understanding the large oval, it is important to know that the ring current belt advances toward the earth during an early epoch of the main phase of geomagnetic storms and retreats during a later epoch (Frank, 1971). This and other observations suggest that the location of the ring current changes from 8 Re for weakest substorms (AE =100 nT) to 4 Re for the most intense substorms (AE = 2000 nT).

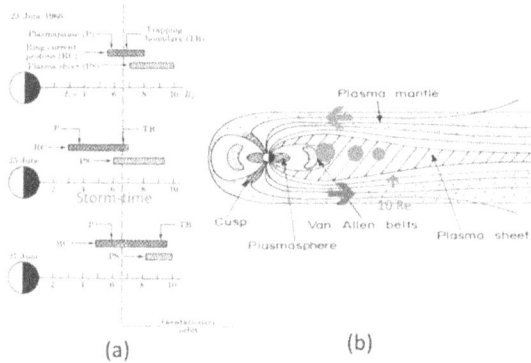

(a)　　　　(b)

Figure 3.10 Advance and retreat of the ring current belt during a geomagnetic storm (Frank, 1971). (b) Schematic illustration of the location of the ring current belt during geomagnetic storms.

Southern auroral oval

There is also an auroral oval in the southern hemisphere. The best location for observing the midday part of the auroral oval in the world is the U.S. South Pole station; it is located at about 75° in magnetic latitude, which coincides with the latitude of the oval in the midday sector.

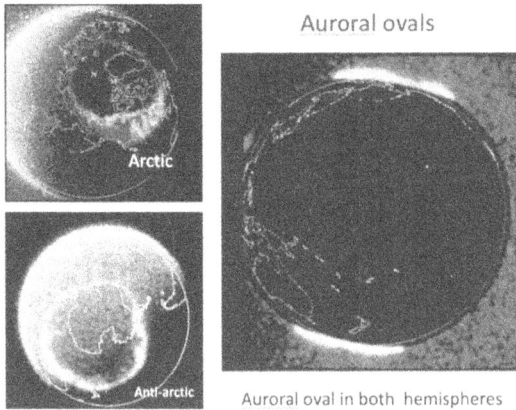

Auroral ovals

Arctic

Anti-arctic

Auroral oval in both hemispheres

Figure 3.11 Auroral oval in both hemispheres (the University of Iowa).

In the early 1970s, I installed a Fairchild all-sky camera there. I assumed I could keep the cover dome at the same temperature as the ones in Alaska. But the South Pole dome froze, so that there was no record for the first year. During the second year, the dome was kept too warm, so that the lens system developed some trouble. As a result, good records were obtained only for the third year on (Akasofu, 1972).

3.5 Diffuse aurora

The auroral oval is surrounded by a diffuse glow with a width of a few hundred kilometers; Figure 3.12. It is called the *diffuse aurora* (Lui and Anger, 1973). The diffuse aurora is usually weaker than the Milky Way, and becomes much brighter during auroral substorms. The diffuse aurora is produced by trapped high energy particles from the radiation belt. This was tested by the NASA airborne expedition, which was coordinated with the planned simultaneous satellite observation (Section 4.3a); the auroral oval is located just poleward of the outer boundary of the outer radiation belt.

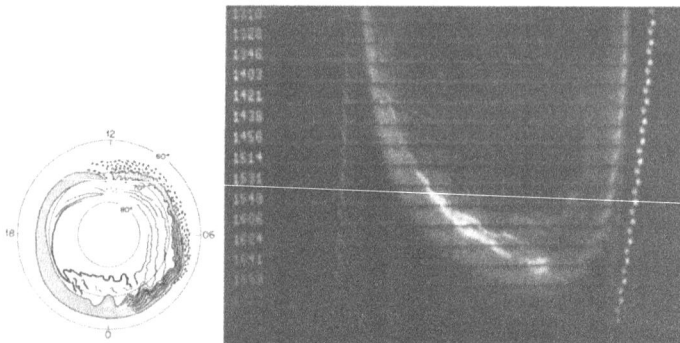

Figure 3.12 Left: Schematic illustration of the diffuse aurora. It appears just outside the auroral oval. Right: The first satellite image of the diffuse aurora (Lui and Anger, 1973).

My motive

The beginning of my study of the auroral oval was simply due to the fact that I had to wait the appearance of the aurora in the northern sky after 9 pm, rather than 6 pm or even earlier in the midwinter. Since the auroral zone was the established fact and all books on the aurora described about the auroral zone. My visual observation did not agree with the concept of the auroral zone.

The auroral oval became a very controversial subject. Even Alfven wrote a paper against the auroral oval. The solution had to wait until an image of the auroral oval obtained by a satellite showed the first image of the oval.

Effects of the X and Y components of the IMF on the shape of the open region

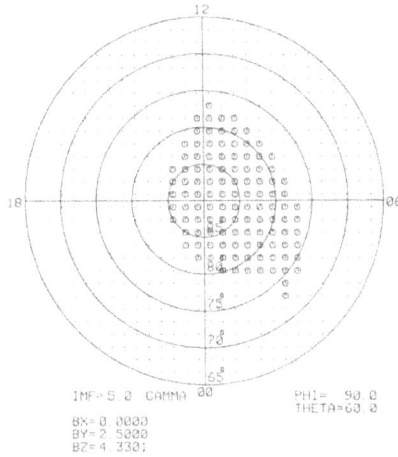

IMF=5.0 GAMMA 00   PHI= 90.0
                       THETA=60.0
BX=0.0000
BY=2.5000
BZ=4.330:

The open field region, when both the Y and -Z component are present; By in this case is + 2.5 nT and Bz = - 4.3 nT. It is known that the auroral oval varys; the morning side tends to become the midnight side.

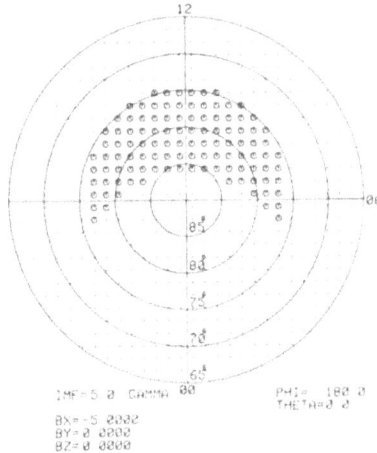

IMF=5.0 GAMMA 00   PHI= 180.0
                       THETA=0.0
BX=-5.0000
BY=0.0000
BZ=0.0000

The open field region when only the Bx component is present. Bx = - 5 nT. It is present in the day side.

Episodes

(1) Emperor and the aurora

I was told earlier that Emperor Showa was very much interested in the aurora.

For the first time during a short talk with him in 1977 at the time when I received the Japan Academy Award, he asked me if the aurora is caused by the solar corona.

During the second time I met with him in 1985, I had the great honor of giving one-hour lecture on the aurora. I sat across a small table from him. At the very beginning of my talk with a slide show, I mentioned that it would not be possible for me to give the talk in terms of polite words; he excused me by nodding.

I was told earlier that Emperor had an opportunity to observe the aurora over Alaska on a polar flight on his way from Japan to meet Queen Elizabeth in London. He observed the aurora soon after the plane left Anchorage (over Fairbanks). Thus, he decided to stay awake through the polar flight to London, but he did not see any aurora during the rest of the flight.

Responding to his question, I showed a satellite image of the auroral oval obtained by Lou Frank, the University of Iowa, pointing out why he did not see the aurora, because he was inside the oval belt during much of the polar flight. He was very satisfied with my explanation.

Among many questions, one of them was if there were any records of sighting of the aurora in Japan (although he was surely aware of it). I told him that there were many records, mostly along the Japan Sea coast (some reported sketches showed even the ray structure).

Then, his question was "How can you be certain about the sighting reports?"

I told him that there is the classification on the sighting credibility: A, B and C. Class A means that the aurora was seen at many places in the world on the same day. He was very satisfied.

Panasonic, a Japanese electronics company, had just completed a video projector for a large screen. It was brought in the Palace by my request.

Emperor watched the video of the first successful *color* auroral video by a joint work by the NHK (the Japanese news company) and the Geophysical Institute of the University of Alaska Fairbanks. I told him that the view on the large screen was just like watching the aurora in Alaska. It was an unforgettable moment.

(2) Trips in Finland and Norway

Finland is the country of Sibelius; I saw his commemorating sign near the Oslo airport. Since they have the well-known Sodankyla magnetic observatory, I visited there twice. Once the researchers there offered me a hot sauna; after it, we jumped into a thinly iced lake together.

I had also an opportunity to visit the Oulu University and found that they are located a little further north (just one mile or two) than the University of Alaska Fairbanks.

The highest latitude university in the world is Tromso University (for which I attended the opening ceremony); I was told that Birkeland set up his auroral observatory there. The second farthest-north is the Oulu University, and the third is the University of Alaska Fairbanks (UAF).

(3) Trip to Huaychulo, Peru

In 1964, with Chapman, T. Nagata and a few other people, I visited the Huaychulo Magnetic Observatory. This observatory was set up to observe the *equatorial electrojet* (the term Chapman coined), a strong electric current in the ionosphere along the magnetic equator (my South Pacific eclipse observation was to see the equatorial electrojet (Section 2.3c).

I found that Peru is a very interesting country. The seashore area is a cold desert where plants get water from the air. Higher up, we encountered tropical trees. We had to go across a hill of about 15,000 feet high before reaching Huaychulo (11,000 feet). We were so tired, that we could barely go up to the second floor of our hotel room.

(4) A. Onwumechili

After I had a chance to observe the equatorial electrojet in the South Pacific Section 2.1), the second chance was at Huaychulo, Peru.

I recall that A. Onwumechili of the Nigeria University was one of the first persons who recognized that an abnormal change of the equatorial electrojet is related to the auroral equatorial electrojet. Nigeria is located under the equatorial electrojet.

I invited Onwumechili to work at the Geophysical Institute in 1972. We found a close relationship between the two jet currents. Apparently, the auroral electrojet enhances the equatorial electrojet through the ionosphere (Onwumechili, Kawasaki and Akasofu, 1973).

On his way back home, I was supposed to pick him up at the university dormitory, but he overslept, so that I had to take him to the airport wearing night gown. A few years later, we met again during an international conference and laughed together while remembering that panic.

(5) Hermann Fritz's compilation of auroral sighting record

At about the time when Elias Loomis established the concept of the auroral zone, Hermann Fritz published a great compilation of the sighting reports of the aurora in 1878.

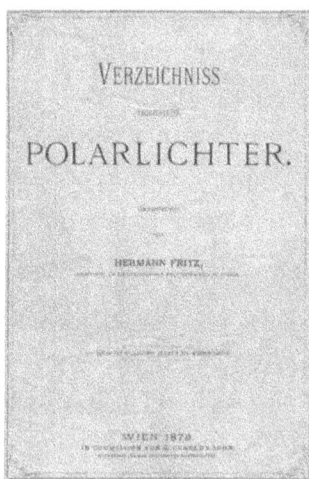

The extensive records of sighting of the aurora by Hermann Fritz (1878).

References

Akasofu, S.-I., Midday auroras at the South Pole during magnetosphere substorms, J. Geophys. Res., **77**, 2,303.

Akasofu, S.-I. and Chapman, S., 1963, The lower limit of latitude (U.S. sector) of northern quiet auroral arcs and its relation to Dst (H), J. Atmospheric and Terr. Phys., **25**, 9.

Akasofu, S.-I., M. Roederer, O. K., Corrick and D. N. Covey, Equatorward shift of the cusp during magnetospheric substorms, *Planet. Space Sci., 29*, 317-320, 1981.

Anger, C.D., Lui, A. T. Y., and Akasofu, S.-I., 1973, Observations of the auroral oval and a westward traveling surge from the ISIS-2 satellite and Alaskan meridian all-sky cameras, J. Geophys. Res., **78**, 3,020.

Armstrong, J. C. and Zumuda, A. J., 1973, Triaxial magnetic mesurements of field-aligned currents at 800 kilometers in the auroral region:Initial results, J. Geophys. Res.,**78**, 6802.

Buchau, J., Whalen, J. A. and Akasofu, S.-I.1970, On the continuity of the auroral oval, J. Geophys. Res., **75**, 7,147.

Feldstein, Y. I., 1963, Some problems concerning the morphology of auroras and magnetic disturbances at high latitudes, Geomagnetism and Aeronomy, **3,** 183.

Frank, L. A., 1971, Relationship of the plasma sheet, the ring current, trapping boundary, and plasmapause near the magnetic equator and local midnight, J. Geophys.Res., **76**, 2265.

Iijma, T. and Potemra, T. A., 1978, Large-scale characteristics of field-aligned currents associated with substorms, J. Geophys. Res., **83**, 399.

Loomis, E., 1860, On the geographic distribution of auroras in the norther hemisphere, Am.er. J. Sci. and Arts, **30**, 89.

Lui, A. T. Y. and Anger, C. D., 1973, A uniform belt of diffuse auroral emission seen by the ISIS-2 scanning photometer, Planet. Space Sci., **21**, 799.

Ness, N. F., 1965, The earth's magnetic tail, J. Geophys. Res., **70**, (13), 2989.

Snyder, A. L. Jr and Akasofu, S.-I., 1972, Observations of the auroral oval by the Alaskan meridian chain of stations, J. Geophys. Res., **77**, 3,419.

Snyder, A.L. Jr. and Akasofu, S.-I., 1974, Major auroral substorm features in the dark sector observed by a USAF DMSP satellite, *Planet. Space Sci.*, *22*, 1,5………..11-1,512.

Chapter 4 Auroral substorms:

Phenomenology

In this chapter, we attempt to establish auroral phenomenology in terms of auroral substorms. We learn first how this concept was established and then how it was confirmed. Before this study, there was a well-established concept, but was not totally consistent with auroral observations based on the IGY all-sky camera network. Fortunately, this subject is now greatly expanded in terms of magnetospheric substorms.

The aurora is not quietly sitting in the sky. During a fairly active night, it becomes suddenly active for one or two hours, covering a large part of the sky. This is one of the most spectacular phenomena in nature.

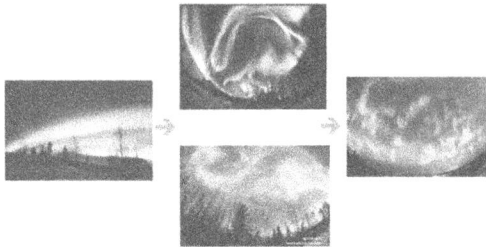

Figure 4.1 Transition of auroral activity, from a quiet condition in the evening sector to active conditions during the midnight sector and patchy display in the morning sector during a fairly active night (GI). Such a view is called the fixed pattern in this book; See also Figure 4.2.

Polar explorers in early days watched auroral displays and described them in splendid ways. Here is a description by polar explorer John Franklin:

"Few nights without a greater or less display of the Aurora Borealis, that wondrous phenomenon whose existence after more than half century of research, is yet unaccounted for satisfactorily. Language is vain in the attempt to describe its ever varying and gorgeous phases; no pen nor pencil can portray its fickle hues, its radiance, and its grandeur."

A study of this phenomenon has become my lifetime project, together with the auroral electrojet, a strong electric current along the auroral oval, causing intense magnetic disturbances (often more than 1000 nT);

It was this magnetic field ('magic hand'), which moved the spotlight at the Onagawa Magnetic Observatory (Section 2.1).

4.1 Fuller's study

During the 1930s, V. F. Fuller and his graduate student E. H. Bramhall at the Department of Physics of the University of Alaska, made a photographic study of auroral activities throughout many nights (Fuller and Bramhall, 1934). It was the period of the Second Polar Year (Section 1.5b).

Based on their study and a study by Jim P. Heppner (1954) (his thesis work at Cal Tech was based on his observations in Fairbanks), it had long been believed that auroras were relatively quiet (quiet arcs) in evening hours, active in midnight hours (violent arc motions) and became patchy (disintegration of arcs, often referred to as "break-up" of auroral arcs).

Figure 4.2 Left: V. F. Fuller and the publication by Fuller and Bramhall (Library of the university of Alaska Fairbanks). Right: Their idea of the fixed pattern of auroral display: quiet in the evening, active in midnight hours and "patches" in morning hours. The earth and an observer on it rotate once day under such a pattern of auroral activities.

Thus, it had long been thought that the earth (and observers upon its surface) rotated under such a *fixed* auroral activity pattern (quiet-active-patches) in the sky once a night. This concept of auroral activity, called here the "fixed pattern," had been believed firmly for a few decades by most auroral researchers until the end of the IGY, because such a case can occur, when the IMF southward component (IMF [-Bz]) happens to occur at the observing location during the midnight hours.

Once, I was asked to examine a large wooden box at the archives of the University of Alaska Library, which had been left by Fuller. It was full of glass photographic plates, notes and correspondence with Carnegie Institute of Washington and also with Carl Stormer (Section 1.4c).

It must have been very hard to operate an old camera in those days on extremely cold nights; one of the best photographs is shown in the cover page of their report (Figure 4.2).

Figure 4.3 shows a little more detail of auroral displays than the right side of Figure 4.2. The auroral display during its activity is different, depending on local (magnetic) time. There are a great variety of displays. The IGY all-sky camera network showed that such a pattern can happen twice or three time in 24 hours, not only once.

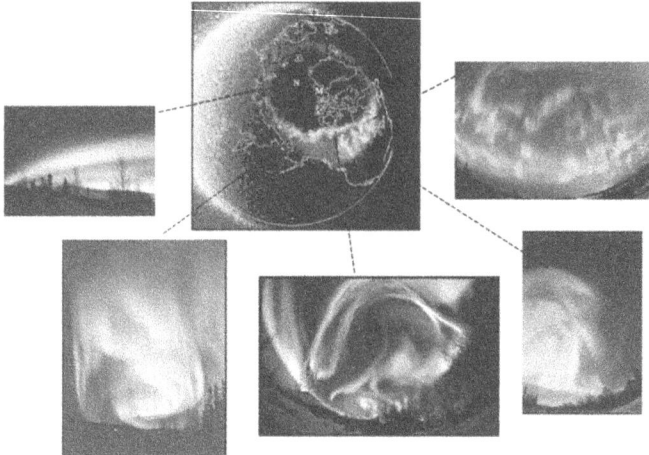

Figure 4.3 Auroral display is different, depending on local (magnetic) time.

4.2 Auroral substorms

When I began to observe the aurora, I made many sketches of auroral displays over the whole sky and compared them with all-sky camera films on the same night. I was greatly surprised that all-sky images were so different from my sketches. The reason I found was that I tended to pay much attention to prominent displays in the sky rather than the auroral conditions over the whole sky. This was the way I learned how to 'read' all-sky camera images. Then, I learned much more by projecting the images on a map, since all-sky camera images are very distorted, much more than a wide range camera images.

Looking back, I took *inadvertently* three steps in conceiving the concept of auroral substorms.

First of all, I named some distinct displays in all-sky films. They are a sudden brightening, poleward advance, surge, patches, omega band and others; the cause of each display is explained in Section 6.7 (as much as we know now);

the term 'patches' had been used before.

The second step was to find if there is some way to organize various displays as a sequence, if they occur together. What should I choose as the beginning of a sequence *(if any)* of displays among such a fascinating variety?

I found that if I choose a sudden brightening of an arc in the midnight sector as the initial display, it is possible to consider systematically a sequence of displays over the whole polar region. The sudden brightening of an arc is one of the *simplest* displays among others (Figure 4.4).

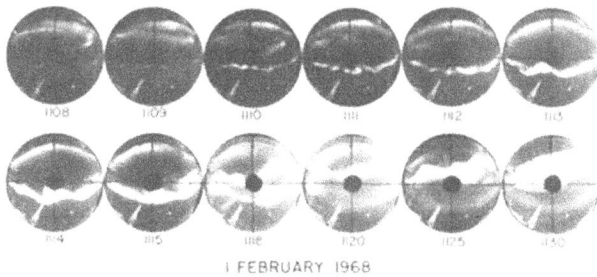

I FEBRUARY 1968

Figure 4.4 An example of the initial brightening of an arc right overhead, signaling onset of the expansion phase. After the initial brightening, arcs advance immediately poleward (GI).

I noticed that the poleward expansion in the midnight sector causes a large wave in the evening sector. Thus, I examined all-sky images at a few Siberian stations and found the wave propagates in the evening sky along the auroral oval. The omega bands and patches tend to occur about 30 minutes after the initial brightening in the morning sector.

It was during the second step, I *happened to* recognize that the daily sequence of quiet-active-patches determined by Fuller (Figure 4.2) occurs twice or even three times in one very active night. This was a great surprise to me.

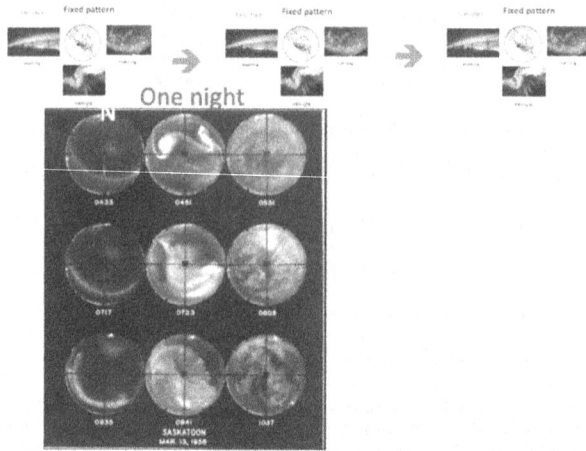

Figure 4.5 A set of all-sky camera images; the pattern of quiet-active-patches occurred (in row) three times in one night. The auroral pattern developed by Fuller repeats three times during active nights (during an intense geomagnetic storm).

Since the earth does not rotate three times a night (momentarily, I thought), this finding gave me an opportunity to examine further *simultaneous* Siberian (in the evening) and Canadian (in the morning) films, when auroras over Fairbanks (in the midnight sector) became active; the IGY all-sky camera network made this work possible. Suddenly, I became more interested in examining auroral activity over the whole polar sky. Thus, the third step began.

In the third step, I repeated a similar examination when Canadian aurora (midnight) became active, simultaneously in Fairbanks (evening) and Scandinavia (morning). Auroral display in the midday side is based on other stations including films from the South Pole station, which is located magnetically at 75° S (the best place to observe the midday sector of the oval).

It was in this way that auroral activities over whole polar sky became gradually clear. It was in this study, *the initial brightening of an arc, which I chose as the initial display of the sequence, was helpful in finding the second or the third step, because auroral activities on an active night are so complex.*

Actually, the three steps were not a systematic approach. I went back and forth many times during three years in reaching the concept of auroral substorms. Perhaps, no one has examined more all-sky films than me.

My view of a typical auroral display over the whole polar sky at that time may be summarized as follows:

The onset of display is initiated by a sudden brightening (or sudden appearance) of an arc at the southern boundary of the auroral oval. The sudden brightening of the arc is then followed by its rapid poleward advance, which causes westward travelling waves along the auroral oval (westward travelling surge [WTS]), the speed of 25 km/s) and disintegration of arcs (break-up) in the morning sector; this results in eastward drifting (300 m/s) patches in the morning sky. The early part of the displays (about one hour) is called the *expansion phase*, and later part the *recovery phase (lasting about 2-3 hours)*.

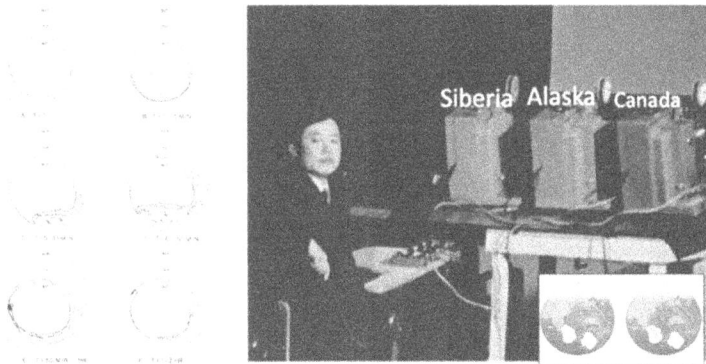

Figure 4.6a Left: View of auroral activities in the whole polar sky based on an analysis of all-sky camera data; the top of each circle is the noon (Akasofu, 1964). Right: Analyzing simultaneous all-sky camera films from Siberian, Alaskan and Canadian statins (GI).

Many other displays, such as torches and the diffuse aurora have been considered later, particularly after DMSP images (see [d]) became available.

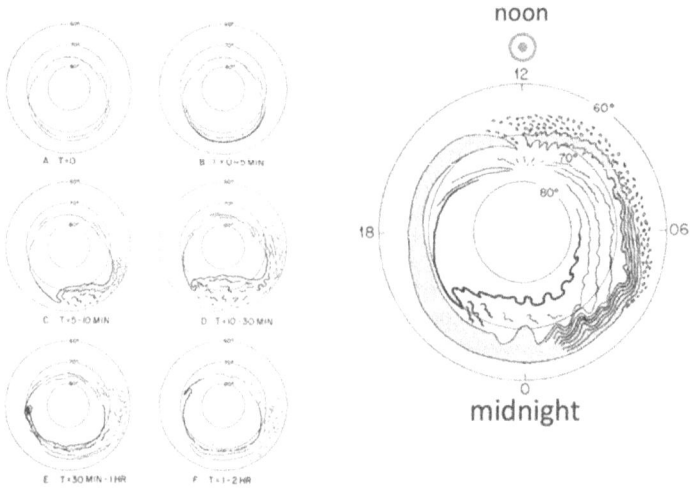

Figure 4.6b Left: The sequence of auroral activity. Right: Some details of the auroral activity at about the maximum epoch.

If I could observe auroral substorms from high above continuously for several days, an auroral substorm occurs, on the average, about 6 times in 24 hours. When one happens to be in the midnight sector (either Siberia, Alaska, Canada or Scandinavia), one can observe the whole display once. If one is not in the midnight sector, one can observe a part of the series of auroral displays, at the beginning, middle or later parts.

During more active periods, it occurs 12 times in 24 hours, thus twice a night, so that one could observe a full substorm activity, if one is located in the dark sector. Thus, the occurrence rate of a significant IMF (-Bz) component (swinging IMF field lines) is, on the average, about 6 times in 24 hours.

Chapman was very happy on my auroral study and suggested the term *"auroral substorm"* for this particular phenomenon.

The reason for the term **sub**storm is that during a major auroral activities (called an auroral *storm*, corresponding to a major geomagnetic *storm;* see Figure 2.23) consist of several auroral **substorms.** Therefore, *an auroral substorm is the unit of auroral storm.*

I sent my substorm paper (Akasofu, 1964) first to Journal of Geophysical

Research, but an editor rejected it. I then sent it to Planetary Space Science, D. R. Bates, the editor, accepted it.

The reason for the rejection was "not useful". At that time, auroral science was limited only to a study of auroral emissions, namely spectroscopy of the aurora. Indeed, when I was scanning all-sky firms, a friend of mine mentioned: "Why are you scanning all-sky films from many stations? The aurora is the same at any place." Certainly, the aurora may be spectroscopically the same, but I was interested in simultaneous auroral displays in different locations in the whole polar region. Much of my analyses of all-sky camera work was supported by the National Science Foundation (NSF).

Feldstein agreed with my concept in general and added many features. Jim Dungey was the first theorist who recognized the significance of auroral substorms.

The 1964 publication of this paper was timely, because soon afterward, many auroral and magnetospheric observations took the concept of auroral substorms as a guide. Many magnetospheric researchers based on satellites asked me what might be happening in the aurora when their satellite observations showed some interesting changes. Indeed, many of their observed changes occurred during auroral substorms. Ching Meng, my first graduate student, participated in this work (Figure 4.7).

My 1964 paper became the most cited paper in space physics for many years and is still quoted; I was listed in "the 1,000 Most-Cited Contemporary Scientists" [Current Contents, 1981]).

The above description of auroral activities occurs during a fairly quiet period. The whole activity becomes much more complex when solar gas clouds called coronal mass ejections collide with the magnetosphere, namely during a major geomagnetic storm; several auroral storms occur in succession. In such occasions, it is not necessarily easy to distinguish three substorms, unless one recognizes the initial brightening.

Figure 4.7 Left: With Ching Meng, analyzing all-sky camera films (IBM). Right: Meng became my first Ph.D. student (GI).

4.3 Proving the concept of auroral substorms

(a) Airborne expedition

Many auroral colleagues had difficulties of accepting the concept of auroral substorms; they firmly believed in Fuller's study. This was understandable, because no one could stay under the midnight sky for about six hours. Further, no one had an opportunity to observe auroral activities from far above the polar region at that time.

In order to convince my colleagues, I asked NASA if I could fly their Galileo plane from the East Coast to Fairbanks. By flying westward, a jet plane could remain under the midnight sky for six hours against the earth's rotation. With me and others aboard, the Galileo plane flew several times what we called the "constant local time flight" in order to confirm that auroral substorms; it could occur twice during the flights in active nights. That is just what we saw.

The U.S. Air Force KC-135 plane (the Flying ionospheric sounder) also participated in this project to confirm the concept of substorms, flying from Boston to Fairbanks many times. Jurgen Buchau was in charge of the flights.

With the NASA Galileo plane, we flew also to many regions in the Arctic. We went to northern Norway to see the midday aurora (the midday part of the auroral oval). Northern Norway is only the location where we can see the midday aurora in the northern hemisphere (further only during the winter solstice time). I recall that Bob H. Eather (1967), spectroscopist from Australia, was excited by watching the midday aurora, saying "The aurora is red!"

We flew around the geomagnetic pole to see any special type of the aurora. One time, the plane lost all the navigation devices during the flight, and I was asked by the pilot to find the high grain tower in Churchill, Canada, by watching the radar screen in the cockpit. Churchill was one of our airborne operation bases (without any guiding device at that time).

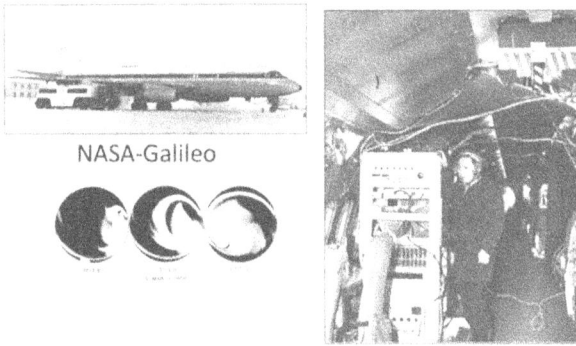

NASA-Galileo

Figure 4.8 Left: NASA Galileo plane (NASA) and a set of all-sky camera images taken from the plane (westward traveling surge). Right: On board the NASA Galileo (NASA).

In one of those flights, we confirmed the location of the initial brightening of an arc with respect to the auroral oval and the Van Allen radiation belt. This observation was crucial in understanding substorm onset (Chapter 6). My analysis of all-sky camera films showed that when the initially brightening arc is located at the southern boundary of the auroral oval, auroral substorms develop most clearly (in the way I described in my 1964 paper). That location is connected to the equatorial plane at about a distance of 6 Re from the earth, not in the magnetotail (say, 20 Re). This was also the confirmation flight of the result shown in Figure 3.4.

This was confirmed by the simultaneous coordinated observation of the aurora and satellite observation of particles during one of the Galileo flights as Figure

4.9 shows.

The precipitation from the outer radiation belt has characteristic gaps during each spin of the satellite and produces the diffuse aurora (Section 3.5). On the other hand, the precipitation in the oval is intense and cause arcs; details of the precipitation in the auroral oval are shown in Figure 1.13.

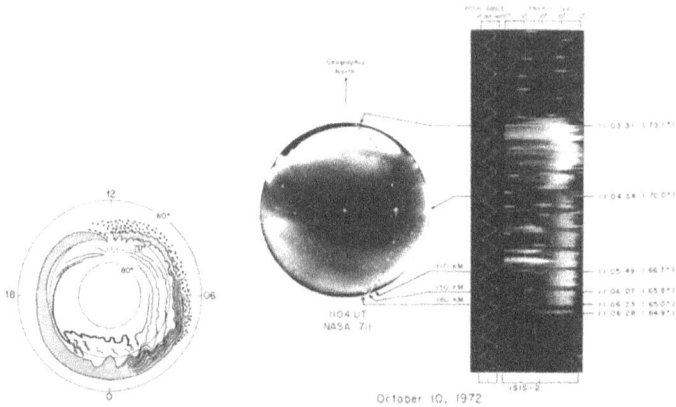

Figure 4.9 A simultaneous observation of the aurora along a trajectory (south to north) of a satellite in the late evening sector. This fight occurred soon after substorm onset. Middle: All-sky image shows both active auroras to the north and the diffuse aurora to the south (NASA Galileo Airborne Expedition). Right:: The simultaneous satellite record of auroral electrons, showing discrete and diffuse precipitations (caused by the trapped particles in the outer radiation belt).

Left of the figure shows a schematic auroral condition at the maximum epoch of the expansion phase. In the evening, the discrete aurora; region and the diffuse aurora are clearly separated.

This was before the Space Shuttle days. NASA took our flights as a good chance to observe how people confined in a small compartment in an extreme condition would behave. Thus, a psychologist was with us. We teased him once by acting odd.

During one of Galileo flights, a staffer at the Geophysical Institute contacted me, saying I should visit Christian Elvey as soon as possible.

Elvey, a former GI director, was in a hospital in Tucson, Arizona. I rushed to

Tucson. He and I sat at the edge of his bed and scanned the NASA films ("constant local time flight") together.

"Syun, you did a good job," he said.

That film helped him to accept the substorm concept. He passed away about a week later.

Unfortunately, the NASA Galileo plane collided with another plane near their base airport. We lost most of the crew with whom we flew on the aurora missions. When I was informed about the news, I could not sleep as I remembered all the pleasures and difficulties we had together during the flights in the Arctic.

I participated also in other KC-135 flights for auroral observations. For this purpose (the training session), I was with fighter pilots. They had a very good lecture series. I learned for the first time how each part of our body works, including eyes and ears. I passed an early examination on physics of the atmosphere, but refused to participate in learning the cockpit ejection system.

In one occasion, we sat in a large room that can be vacuumed, so that we could experience a low atmospheric pressure (high altitude) situation. I saw a balloon hanged from the ceiling began to expand. I saw red and blue spots flying around me, and I lost the ability to count $(1 + 2)$.

(b) First satellite observation from high above

All these efforts were prior to the days of satellite located high above. We needed the confirmation of the concept of auroral substorm by a satellite a few earth's radii above the polar region for an extended period.

Thus, I was delighted to see the first image of the auroral substorms by a Canadian satellite in 1972. It was very similar to what I imagined based on a large number of all-sky camera images; Figure 4.10.

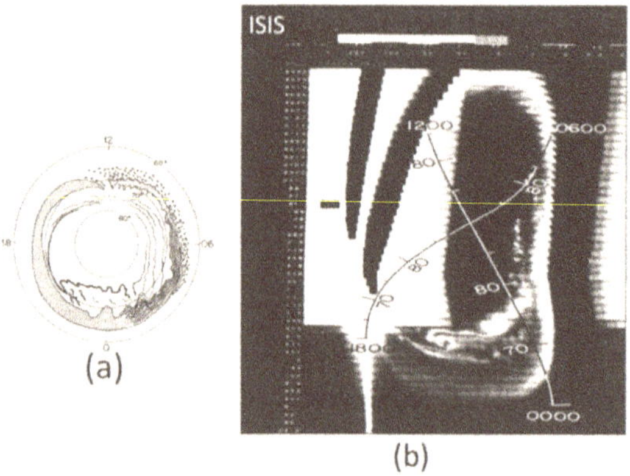

8

Figure 4.10 (a) Sketch of auroral activities during the maximum epoch of auroral substorms. (b) The first satellite image of auroral substorms by a Canadian satellite; note the westward traveling surge in the evening sector (Cliff Anger).

(c) Confirmation from high above the north pole by a satellite

Finally, the real chance of the confirmation of the concept of auroral substorms came. A series of satellite images from high above the north polar region (about 3 Re above) was successfully obtained by Lou A. Frank at the University of Iowa in 1982 (Frank et al., 1982). It took 18 years to confirm solidly my all-sky camera analysis by a satellite after the publication of my first paper in 1964.

I was waiting for the arrival of the first images with Lou at the University of Iowa; I congratulated Lou for his great success for this spectacular imaging, and he congratulated me about the accuracy of my all-sky camera analysis in 1964. We had lunch together and recalled various difficulties we encountered during in our research.

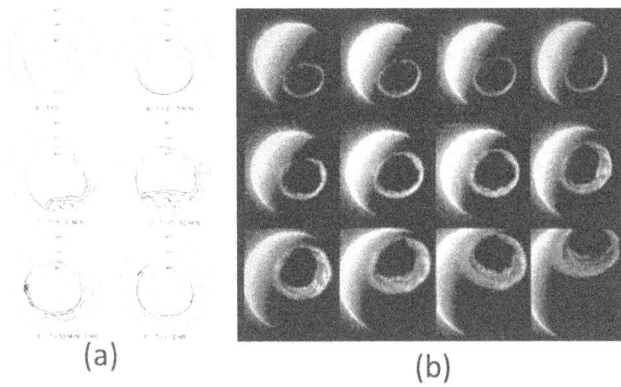

(a)

(b)

Figure 4.11 Comparison of my analysis results by all-sky camera analysis in 1964. (a) It is the same as Figure 6.4a. (b) The first satellite images of an auroral substorm taken from above the norther polar region in 1982 (L. A. Frank).

During many auroral conferences, Lou often showed first my substorm pattern in 1964 before his many fascinating satellite images. I appreciated his support of the substorm concept. My auroral substorm pattern was simple enough to help understanding excellent but complicated satellite images for the audience.

Later, many satellites took a full set of images of auroral substorms, from the onset of the expansion phase to the recovery phase.

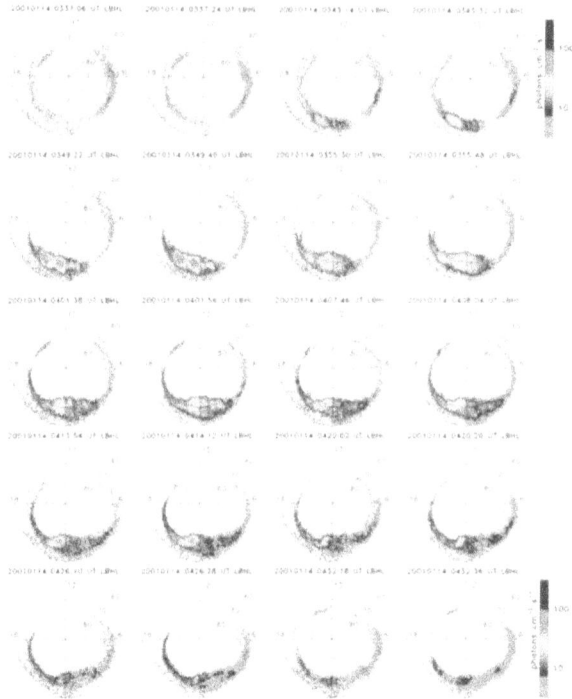

Figure 4.12 The full development of an auroral substorm observed by a satellite in the geomagnetic coordinate (C. G. Parks)

(d) DMSP satellites

The U.S. Air Force had a very extensive satellite observation of weather in the polar region during the Cold War from 1970-1980 and later. It was called the Defense Meteorological Satellite Program (DMSP). In its earliest days, their image showed 'mysterious' lights in Siberia. They must have thought that the Soviet Union might have been developing a secret device. I was asked by their Boston Laboratory of the Air Force to make sure that the lights in the top-secret images were the aurora.

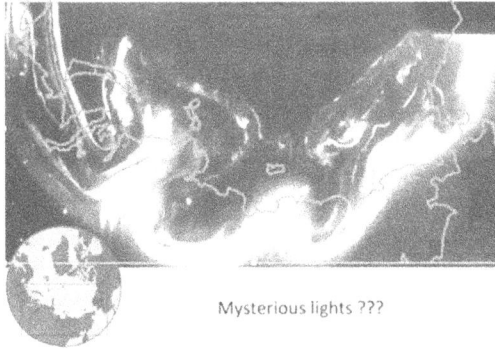

Mysterious lights ???

Figure 4.13 An example of DMSP images over Siberia, showing auroral activities (USAF).

The DMSP images helped us greatly in advancing auroral morphology by its high space resolution and the wide covered area (about one half of the polar region, which is equivalent to more than 20 simultaneous all-sky images. Although the images were taken every 90 minutes (orbital period of the satellite), these images confirmed my pattern of auroral displays in general (Snyder and Akasofu, 1974). We learned much more auroral displays in larger scale than all-sky images. Col. Lee Snyder, my former gradient student, helped us release of many of the images for scientific use after their use. Don Kimball located DMSP images on geographic maps).

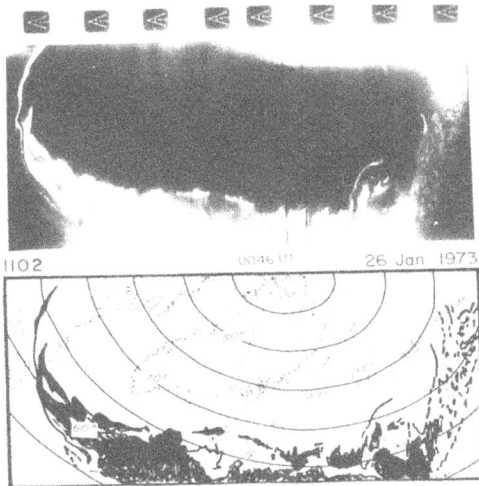

Figure 4.14 Upper: Typical example of DMSP image (nightside) at about the maximum epoch of auroral substorm (DMSP). Lower: The upper image is projected on a map.

105

4.4 Magnetospheric substorms

During the early period magnetospheric physics, a number of colleagues found that there occur a variety of magnetospheric and ground-based phenomena in association with auroral substorms. Thus, the concept of **magnetospheric substorms** was soon developed and established in helping to organize many polar and satellite phenomena. The concept of auroral substorms became very useful in studying simultaneously polar ground and magnetospheric observations. We attempt to synthesize these observations in Chapters 5 and 6.

As a result, the following subjects have been studied:

Auroral substorms

Polar magnetic substorms (cf. Kamide et al. 1982)

Ionospheric substorms (cf. F. Tom Berkey, 1968)

Proton auroral substorms (cf. Bob H. Eather, 1967)

X-ray auroral substorms (cf. Winckler et al., 1959)

VLF and ULF substorms (cf. T. S. Jorgensen)

Various magnetic observations by satellites

For the early development of the concept of magnetospheric substorms, see Akasofu (1968).

My motive

When I began to work at the Geophysical Institute, I did not know the auroral recording instrument called 'all-sky camera'. I noticed that the Photo Department was developing a large number of 16 mm films, and I wanted to know what they were. I was fascinated at once by the all-sky films when I could scan them.

Then, I compared them with my sketches of the aurora and learned that all-sky cameras record the aurora all over the sky correctly (although the images are greatly distorted).

This scanning of all-sky camera films was only to satisfy my curiosity. I had no idea at that time that my study of auroral activity over the whole polar sky would become one of the basic concepts in space physics (auroral substorms, magnetospheric substorms).

When I began to scan all-sky films recorded in Fairbanks, I noticed that there is something is wrong about Fuller's study. I noticed that the well-established concept established by Fuller (quiet in the evening, active in midnight hours and broken-up patches in the morning) can occur twice or even three times in one active night. This fact was confusing me, because Fuller's concept, as I learned, had been so firmly believed at that time.

Finding such cases prompted me to examine the *simultaneous* all-sky images in Siberia (evening) and Canada (morning), when the aurora becomes suddenly active over Alaska.

I was simply surprised that my findings have become so useful by others. I had no intension or motivation of establishing a new morphological theory at that time. It so happened that many researchers found later that their observations (both ground-based and satellite-based) can be organized well on the basis of the concept of auroral substorms. They established the concept of *magnetospheric* substorms.

Episodes

(1) International Conferences of Substorms (ICS)

Auroral researchers organized a series of international conferences, called *the International Conference of Substorms (ICS)*.

Joe R. Kan, a GI space physicist, was the first to propose such a series of the conferences. The first one was held in Kiruna, Sweden in March 1992. It was held every few years and was attended by more than 100 researchers from many countries. The third one was held at the University of Alaska Fairbanks.

It so happened that great auroral activities occurred during the conference night, so that many participants witnessed a typical auroral substorm for the first time.

In another auroral conference in 2010 in Fairbanks, we found that about 10 participants among 120 had not seen the aurora before, and had a good chance of observing a typical auroral substorm during the conference. The organizer of the conference asked me to provide a "certificate" to them, stating that I certify their auroral observation. I am very glad to know the ICS is continuing today.

Left: The opening ceremony of the first International Conference of Substorms (ICS) in Kiruna, Sweden. Joe Kan made the opening speech, March 23, 1992 (the Kiruna Observatory). Right: Collection of the abstracts of papers presented at the Conference.

International Conference of auroral substorm (ICS) held at the University of Alaska Fairbanks in 1984 (GI).

(2) Publication of the first book on auroral substorms

In 1967, I published the first book on auroral substorms titled *"Polar and Magnetospheric Substorms."*

I dedicated it: "To Sydney Chapman who, *unbeknown* to most readers, has encouraged and inspired the world's magnetic and auroral observatories to maintain the essential records upon which our understanding of geomagnetism and the aurora rests." Chapman, a theorist, visited many magnetic observatories in the world. Feldstein published the Russian translation of my book soon afterward; I learned how my name is spelled in Russian.

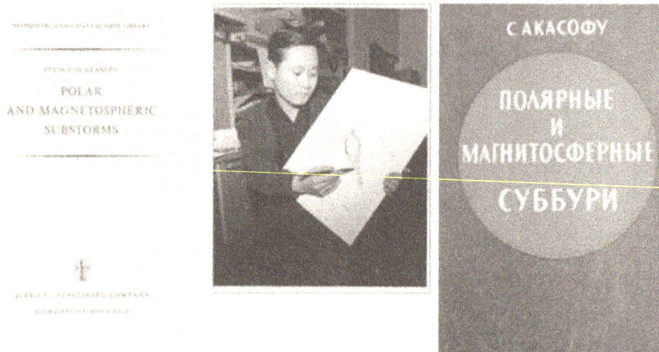

Left: My first book titled "Polar and Magnetospheric Substorm", published in 1967. It was dedicated to Sydney Chapman. Middle: I was explaining the content of the book (GI). Right: Russian translation of my book by Yasha Feldstein.

(3) My visit to China

My Chinese colleagues invited me to talk on auroral substorms in the early 1970s. In China, there were many ancient auroral observations, including many sketches. Although I cannot speak Chinese, I could write some Chinese characters on the blackboard in explaining some features of auroral displays.

At that time, I could see only one neon sign in Beijing in the evening: "Beijing Duck." We went to watch a performance of the Red Army. It was the time when the Cultural Revolution was still in progress.

During the second visit, I wanted to take 20 packages of Sapporo ramen for my dinner (I did not favor some Chinese dishes). By some mistake, I ended up taking 200 packages. So, I invited about 20 colleagues every evening for dinner. They were very happy to be together; we discussed the aurora in my hotel room.

We also visited the Great Wall together. Since then, I had an opportunity to visit China a few more times for conferences. Their economic development was astonishing.

(a) Lecturing at the Beijing University. (b) Visiting the Great Walls.

(4) My visit to the Kiruna auroral observatory

I often visited Kiruna, Sweden. The Kiruna Auroral Observatory was established by Bengt Hultqvist; the observatory was a sort of sister relationship with the Geophysical Institute in auroral studies. It is now developed into the Space Physics Institute of Sweden. Bengt and I had many good discussions on the aurora.

With Bengt Hultqvist (Kiruna Auroral Observatory). He established the Kiruna auroral Observatory.

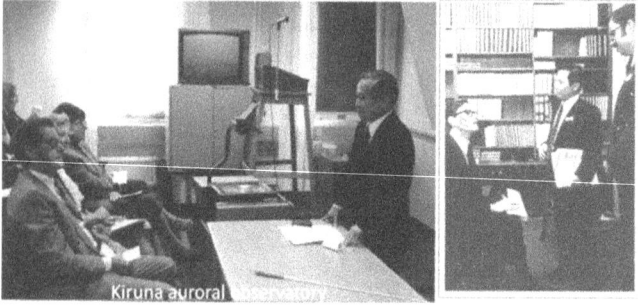

Left: Seminar at the Auroral Observatory in Kiruna, Sweden. Director Bengt Hultqvist on the left side (his wife in the middle) was a close friend of mine (the Kiruna Observatory). Right: Discussion with Charlie P. Kennel and Vatanus M. Vasyliunas after my talk at the Kiruna Observatory, Sweden (the Kiruna Observatory).

(5) Leiv Harang and my visit to Norway

One of the unforgettable pioneers of auroral research was Leiv Harang. He succeeded Birkeland and established the auroral observatory at Tromso in Norway. He worked also on the ionosphere with W. Stoffregen. When I was invited to his home near the Oslo Harbor, he told me about the early history of auroral research in Norway and his ionospheric research during the WW II.

After dinner, the Russian invasion of Eastern Europe came as a news by radio. I cannot forget the moment when he and his wife were listening the radio. When I was attracted by a beautiful candle holder on a desk, his wife offered it to me. On another occasion, Stoffregen took me to a lake and had lunch together on his boat.

Left: Leiv Harang (the Tromso University). Right: Candle holder given to me by Mrs. Harang.

Alv Egeland invited me to contribute to his "The Kristian Birkeland Lecture Series." The title of my lecture was "Heliomagnetism." Egeland took me to Stormer's old office and showed me his articles and notes. With several colleagues of auroral science, we went to swim in the Oslo harbor together after a conference.

Figure 4.22 With Alv Egeland at the Oslo University (Oslo University). Egeland was a very active auroral physicist.

(6) Richard Feynman's visit

In 1982, a group of undergraduate students of the University of Alaska Fairbanks invited Richard Feynman, Nobel Prize winner in physics in 1965, but I did not know about his visit.

However, he suddenly showed up in my office and asked me about the aurora. After asking me about my work, he said he wanted to work on the aurora, but he said he had to get the permission from his sister, Jo Ann Feynman, since she had asked him not to work on the aurora.

In 1983, there was a conference on *Magnetospheric Currents* (organized by Tom A. Potemura, 1984). There I met Jo Ann, who told to me with her stern voice "You are a trouble maker !" At first, I had no idea what she was talking about. Obviously, she did not allow Feynman to work on the aurora.

(7) Russian trips

During my first trip to meet Yasha Feldstein in 1968, I had a rare opportunity to go deep into the Kremlin and visit Lenin's simple living quarters, with a few pots and pans. I sat on his chair. His bookshelf had a book by a Japanese communist.

Yasha and I watched the ballet Sleeping Beauty in Petersburg. An image of the aurora was shown in the background of the theater.

On that trip, I took Pan American Airways. On my return trip, I went to the Moscow airport. There, a soldier with a gun asked me to go to an Aeroflot plane without listening my protest as I showed him my Pan American ticket. On the plane, I was not sure in which direction the plane was flying, Siberia or Europe. I was so relieved when I saw the sign of the Copenhagen Airport.

During another time (Gorbachev's time), I was invited for a dinner by my Russian colleague in his apartment. During the dinner, there came the sound of hammer from his neighbor. Immediately, all Russian colleagues stood up, tapping the table and shouting "Perestroika!"

It was an easier time for communication with Russian scientists than before. They often sang together Russian folk songs after conferences. Professor A. Lebedinsky told me his study of the aurora in his early days.

Moscow

Petersburg

Left: A snapshot of Moscow, when I visited for the first time. Right: Yasha Feldstein and I watched the ballet "Sleeping Beauty" in Petersburg. The auroral painting was in the background.

From the left Professor S. N. Vernov, O. V. Khorosheeva and Yasha Feldstein at the Moscow University. Professor Vernov found the radiation belts a little later than Van Allen (Moscow University).

(8) Substorm conference at Los Alamos

There have been several conferences on auroral substorm and magnetospheric substorms in the world, one in Loa Alamos organized by Ed Hones. There were many intense discussions among attendants, and was generally very friendly conference.

With Carl E. McIlwain, Jack R. Winckler (with cap) and Ching-I. Meng, dancing after lunch time during the Los Alamos conference (Los Alamos).

(9) Auroral kilometric radiations

Once, on my way back from the Air Force laboratory in Boston (receiving DMSP images), I made a short visit at the University of Iowa and met Don A. Gurnett. He showed me an interesting radio emission data from a satellite (located above the ionosphere) and wondered if they were related to the aurora.

It so happed that the period of his data coincided with that of DMSP data, which I received in Boston. It was very clear that the radio emission occurred when the aurora was active; Don called the emission the *auroral kilometric emission* (Voots et al.,1977).

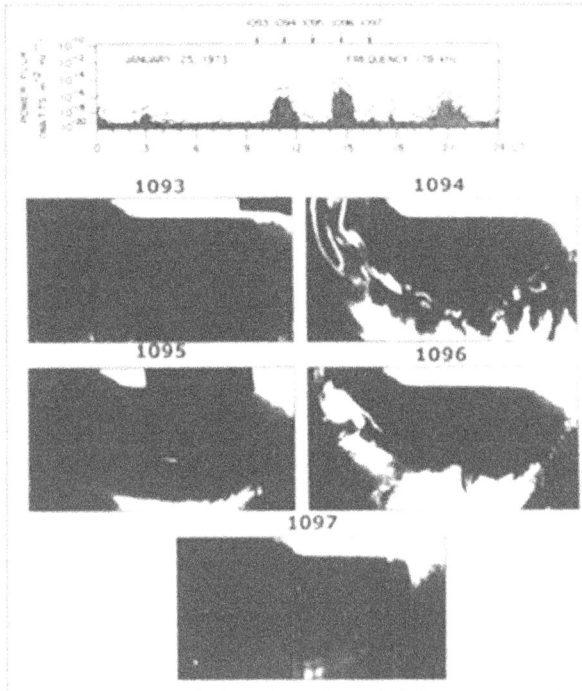

Above: Auroral kilometric radiation record (D. Gurnett). Below: DMSP images corresponding to the time of the auroral kilometric radiation.

References

Akasofu, S.-I., 1964, The development of the auroral substorm, Plant. Space Sci.,**12**, 273.

Akasofu, S.-I., _Polar and Magnetospheric Substorms_, 1968, D. Reidel Pub. Co., Dordrecht, Holland, 1968. (also a Russian edition)

Berkey, F.T., 1968, Coordinated measurements of auroral absorption and luminosity using the narrow bam technique, J.Geophys.,**73**, 319.

Buchau, J., Whalen, J. A. and Akasofu, S.-I. 1q970, On the continuity of the auroral oval, _J. Geophys. Res.,_ 75, 7,147-7,160.

Eather, R. H.,1967, Auroral precipitation and hydrogen emission, Phys. Rev., **84**, 203.

Fuller, V. F. and Bramhall, E., 1934, *Auroral research at the University of Alaska 1933-1934*, University of Alaska Press.

Heppner, J. P., 1954, Time sequences and space relations in auroral activity during magnetic bays at College, Alaska, J. Geophys. Res., **59**, 329.

Jorgensen, T. S., 1968, Morphology of VLF hiss zone and their correlation with particle precipitation events, J. Geophs. Res., **71**, 1367.

Frank, L. A., Craven, J. J., Burch, L. and Winninham, D. J., 1982, Polar view of the Earth with Dynamic Explorer, Geophys. Res. Lette. **9**(9),1001, https://doi. org./10.1029/GL009i009p01001

Potemra, T.A. (ed), 1984, *Magnetospheric Currents*, Geophysical Monograph 28, AGU, Washingto, DC.

Ness, N. F., 1965, The earth's magnetic tail, J. Geophys. Res., **70**, (13), 2989.

Voots, G.R., D.A. Gurnett, D. A. and Akasofu, S.-I., 1977, Auroral kilometric radiation as an indicator of auroral magnetic disturbances, J.Geophs. Res., **82**, 2,259.

Snyder, A.L. and Akasofu, S.-I. 1972), Observations of the auroral oval by the Alaskan meridian chain of stations, *J. Geophys. Res.,* **77**, 3,419.

Winckler, J. R., Peterson, L. Hoffman, R. and Arnold, R., 1959, Auroral X-rays, cosmic rays and related phenomena during the storm of Feburay 10-11, 1958, J. Geophys. Res. **64**, 597.

Chapter 5 Auroral substorms: Power and electric currents

⊸⊷⊳⊷⊡⊷✳⊷⊡⊶⊲⊶⊸

The aurora is basically a magnificent *electrical discharge* phenomenon in space around the earth. Thus, it is essential to study *electric current* in understanding auroral substorms, namely how the current is generated by a dynamo process, how it is transmitted for the discharge location and how the discharge can cause the aurora and its activity, auroral substorms. However, this concept of electric current is almost missing in space physics and also in solar physics even today. This particular reason is discussed in the next chapter in association with solar flares (section 7.1 [vi]).

In this chapter, we consider auroral phenomena, as well as in Chapter 6, in terms of electrical *circuit* system, namely the magnetosphere-ionosphere coupling *circuit*. Thus, the first step is how the circuit is powered and then how electric current flows in the connected electrical circuit. The final stage is the discharge in the ionosphere, where the power is dissipated. Auroral substorms are the manifestation of the power dissipation.

This work based on electric currents requires an extensive ground-based observation of ionospheric currents as we discuss in this chapter. In fact, this work distinguishes our method from other studies.

5.1 Electrical discharge in a high vacuum

The public in general had believed for a long time (more than one hundred years) that the aurora was caused by charged particles from the sun, which were captured by the earth's magnetic field and excite the polar atmosphere. This is the knowledge of many people even today. Auroral scientists are responsible for their out-of-date knowledge, not being able to convey our present knowledge.

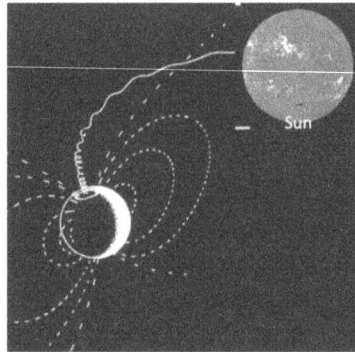

Figure 5.1 Past common belief on the cause of the aurora by the public in general. The auroras (northern lights) were supposed to be caused by charged particles, which were attracted by the earth's magnetic field and entered in the polar region.

Auroral science has considerably advanced during the last 100 years, particularly after 1960, when the space age began. As a result, we can say now that *the aurora is an electrical discharge phenomenon in space around the earth, namely in the magnetosphere.*

A very close discharge phenomenon we can observe is a neon sign, which is familiar with most people. In fact, in terms of basic physics, auroral electrical discharge itself is like the electrical discharge within a neon sign.

A neon sign is made by putting a small amount of neon gas in a vacuumed thin glass tube and by connecting it to a high voltage supply.

This means that *electric power is needed for the aurora* (Lemstrom's experiment, Figure 1.21). Thus, the first questions on the aurora are: *what kind of dynamo (generator) do we have and where is the auroral dynamo located in the sky and further how much power does it generate?* In explaining the aurora in terms of the magnetic field line approach, people hesitate to ask even simple questions on the aurora.

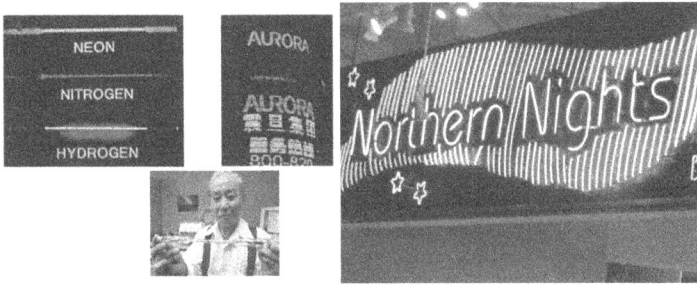

Figure 5.2 Left: Neon signs. Explaining how to make a neon sign. Right: Neon sign in Shanghai and at a shop in Fairbanks, Alaska.

5.2 Auroral dynamo

Electric power is generated by a dynamo (generator), which consists of a coil that rotates in a magnetic field (a magnet). The earth is a gigantic spherical magnet; Section 2.1. The solar wind (a continuous flow of plasma from the sun) is an electrical conductor (consisting of protons and electrons) and plays the role of a coil. Instead of rotating (moving) the coil, the solar wind *blows through the linked magnetic field lines* between the solar and magnetospheric field lines; the solar magnetic field (stretched out by the solar wind; see Figure 2.16).The stretched solar magnetic field is called *the interplanetary magnetic field (IMF)*, in which its southward component (IMF [-Bz]) is found to be the "unknown" factor in my finding (Sector 2.3).

Therefore, the auroral dynamo requires three elements — the solar wind, earth's magnetic field and the interplanetary magnetic field (IMF). Among the three factors, the IMF is most variable (like swinging a rope connected to the sun).

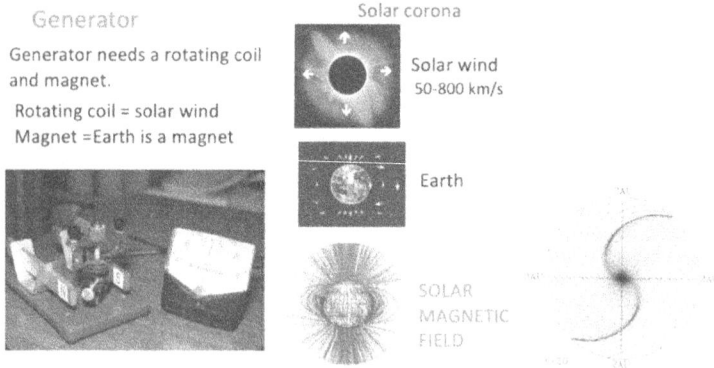

Generator

Generator needs a rotating coil and magnet.

Rotating coil = solar wind
Magnet =Earth is a magnet

Solar corona

Solar wind
50-800 km/s

Earth

SOLAR
MAGNETIC
FIELD

Figure 5.3 Explanation of the solar wind-magnetosphere dynamo. It explains how a dynamo (generator) works. Left: The coil (the solar wind) is rotated in a magnet (the linked magnetic field between the interplanetary magnetic field and the earth's magnetic field/magnetospheric magnetic field). Middle: The three factors for our dynamo, the solar wind, the earth's magnetic field and the solar magnetic field. Right: Solar magnetic field (the interplanetary magnetic field (IMF), stretched out to a 2 au (equatorial plane).

It took more than 15 years for me to find how the above three factors together constitute the auroral dynamo and *how to prove it quantitatively* (Akasofu, 1981) after Dungey suggested that the southward component of the IMF, (-Bz), is responsible for the main phase of geomagnetic storms (Section 2.3).

Since the magnetic field lines has a very large-scale structure, say 0.1 au [1 au = the distance between the sun and the earth] (Section 7.1 [vii]), it may be considered to be a uniform field in scale in the vicinity of the earth and the magnetosphere, oriented in all directions as the field lines swing. Here, we consider the case, in which the IMF is southward-oriented; for other directions, see after Section 3.5.

First of all, the southward component of the IMF (-Bz) of the solar magnetic field lines link with those of the magnetosphere. The linkage points on the surface of the magnetosphere (called the magnetopause) are distributed from the *front side* of the magnetosphere; Figure 5.4; see also Section 3.3.

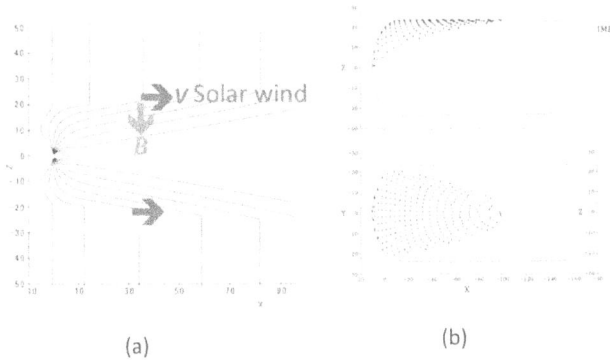

(a)   (b)

Figure 5.4 (a)The solar magnetic field lines carried out by the solar wind link with the earth/magnetospheric field lines. The solar wind blows through the linked magnetic field lines. As a result, a dynamo process occurs (V x B), providing the electric power for the aurora. (b) The linking points of the solar magnetic field line on the boundary of the magnetosphere (magnetopause), both side and upper views)

The power of the auroral dynamo is generated by the solar wind (of speed V), which blows through the linked field lines (B), a dynamo process (V x B); Figure 5.4a and Figure 5.5. Thus, *the auroral dynamo is located on the boundary surface of the magnetosphere (magnetopause)*; the dynamo process occurs from the front of the magnetopause as the distribution of the linked points shows (Figure 5.4b).

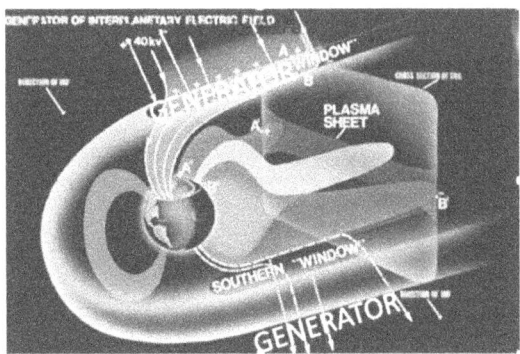

Figure 5.5 Auroral dynamo and the internal structure of the magnetosphere (without the ring current belt); (NASA).

In terms of the magnetic field line approach, the solar magnetic field lines link

123

with magnetic field lines at the front side of the magnetosphere, and the linked field lines are carried to the magnetotail by the solar wind. Then, magnetic reconnection in the magnetotail provides the power.

During 1970s, a number of researchers were trying to determine the relationship between the geomagnetic index AE index with the solar wind parameters, such as pressure p, number density n, electric field E, magnetic field B and their combinations.

Since I learned the importance of the importance of the IMF Bz component, not p or n (Section 4.1), I chose only a combination of V and B, not including p or n (see Vasyliunas et al, 1982), and found that the combination of V and B, namely VB^2 and finally $VB^2\sin^4(\theta/2)$ correlates best with the AE index (Perreault and Akasofu, 1978). This finding led to the concept of a dynamo process. Thus, my analysis of the power is given by $\varepsilon = VB^2\sin^4(\theta/2)S = P$.

In the following two sections, I describe the power of the dynamo and the magnetospheric circuit. We could confirm quantitatively the concept of the dynamo in terms of the basic electric power equation.

5.3 Power of the auroral dynamo

$$P = \int (\boldsymbol{E} \times \boldsymbol{B}) \cdot d\boldsymbol{S} = V\sin^4(\theta/2)\,(B^2/8\pi)S$$

For the solar wind speed $V = 500$ km/s, IMF $B = 10^{-4}$ G (=10 nT), S *(cross-section of the magnetosphere)* $= (l^2\pi)$, $l = 10$ Re, (Re= the earth's radius), $\theta =$ the polar angle of the IMF [the angle $\theta = 180°$ indicates the IMF is oriented southward], the power is:

$$P = 2.0 \times 10^{12} \text{ W } (2.0 \times 10^{19} \text{ erg/s}).$$

This value is comparable with the Joule dissipation rate (90 %, Section 5.8).

Observationally, we could confirm the power ε (= P) by estimating the total dissipation rate during a geomagnetic storm UT (based on Dst and AE indices; these values are converted empirically to physical quantities).

The power ε $(= P)$ and UT have similar time variations (Figure 5.6), while the kinetic energy flux $K = ([1/2]mnV^2S/s)$ shows an entirely different time variations (this was inferred early by the "unknown" factor in Section 4.1); Akasofu (!981).

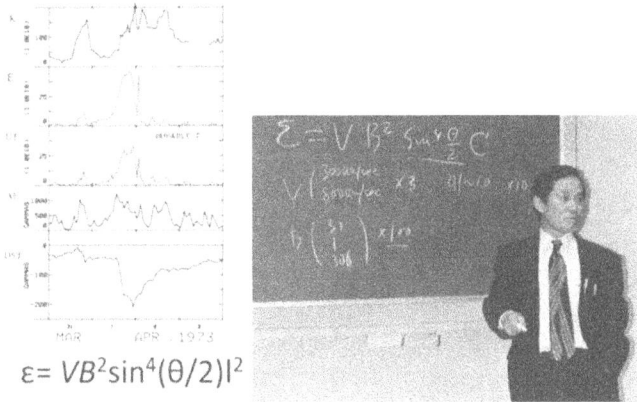

$$\varepsilon = VB^2\sin^4(\theta/2)l^2$$

Figure 5.6 Left: From the top, the solar wind pressure K (= [1/2]mnv²s/s), the power ε (= P) estimated by the solar wind data and the estimated the total dissipation rate UT during a magnetic storm (estimated on the basis the Dst and AE indices by empirically converting). Thee is a good similarity between the dynamo power ε and UT (the magnitude difference is caused by uncertainty of the lifetime of ring current particles; see Section 2.3c), while the kinetic pressure K has no clear relation to UT (Akasofu, 1981). Right: Lecturing about the dynamo process (GI).

The voltage ϕ of the dynamo is about 100 KV; this was measured by satellites by flying across the polar cap; Figure 5.7.

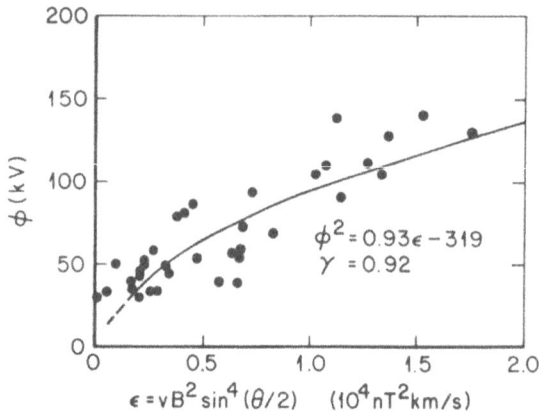

Figure 5.7 The relationship between the polar cap potential ϕ and the power ε (Rice University).

In space physics. the source of the necessary power has almost exclusively been discussed in terms of hypothetical magnetic reconnection, which has not yet been confirmed (Section 7.1 [vi]). Here, we have a solid power process based on the *observed quantities* (Akasofu, 1981).

In this study of electric current and magnetic variations, the intensity of auroral activities is customarily measured in terms of the AE and other indices (based on magnetic records from the auroral oval and other indices). It is important to know the relationship between the AE index and the power, for example, at what level of the power the AE index responds to increasing power ε. For this purpose, the polar cap index is made by magnetic records from several the polar cap stations (excluding the records from the AE stations); this is because the root points of the linked magnetic field lines are concentrated in the highest latitudes during a quiet time (Figure 3.7 [c]), and thus the polar cap index sensitively responds to the power ε (= P) down to 10^9 W (10^{16} erg/s). On the other hand, the AE index responds to the power, only when it rises above about 10^{11} W (10^{18} erg/s).; Figure 5.8.

In this and the next chapter, this quantity 10^{11} W is often referred to indicate that it is the power by which auroral substorms (the growth phase) begins, rather than when the IMF (-Bz) occurs.

126

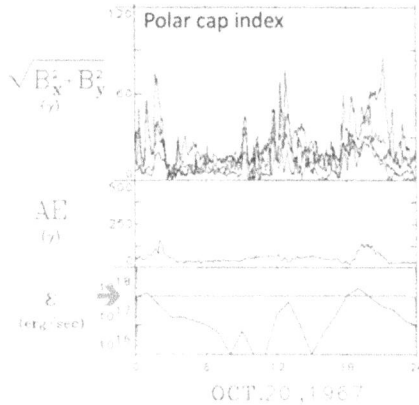

Figure 5.8 From the top, the polar cap index (combined records from highest latitude stations), the AE index and power ε (= *P*); Akasofu (1985). The figure shows that the magnetosphere (the primary M-I system) responds to the power above about 10^{11} W (10^{18} erg/s); the AE index responds to the power above 10^{11} w; see the red line; $\gamma = nT$.

5.4 Circuit: Primary Magnetosphere-ionosphere coupling system

The current system connected directly to the dynamo is called the *primary Magnetosphere-Ionosphere coupling system* (in short, the primary M-I system). In Figure 5.9a, we show the primary Magnetosphere-Ionosphere coupling system. The terminals of the dynamo are located on the equatorial plane of the magnetopause, the positive terminal on the dawn side and the negative terminal on the dusk side. The equivalent circuit (its simplest version) is shown in Figures 5.10 and 6.2.

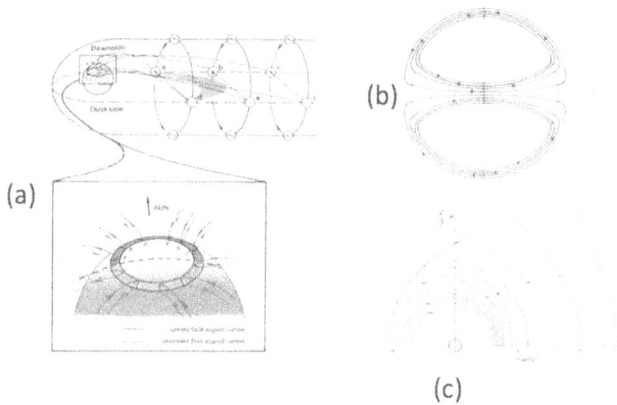

Figure 5.9 (a) The electric current circuits of the magnetosphere, powered by the solar wind-magnetosphere interaction (dynamo); (a) The circuit of the *primary M-I coupling system*. (b) The cross-section of the magnetosphere, which is consisted of two half solenoidal currents. (c) The two half solenoidal currents flow jointly across the equatorial plane (Olson,1984)

The primary M-I system consists of two parts. The first one connects the terminal with the ionosphere by the magnetic field lines (field-aligned currents), as established by many earlier studies in the 1970s and 1980s (cf. Potemra,1984). The field-aligned current intensity is about 10 mA/m^2. When the power is less than 10^{11} W, the electrons carrying the field-aligned currents do not seem to generate the double layer (Section 6.4), so that the double layer cannot accelerate electrons to excite ionospheric atoms and molecules. This fact is important in understanding the growth phase of auroral substorms; Section 6.2).

The second circuit consists of two half-solenoidal currents, one in each hemisphere; the two (north and south) half-circular parts of the currents on the magnetopause are where the dynamo (V x B) process operates (Figures 5.4 and 5.5); note again that the dynamo process occurs from the front side of the magnetopause, so that the two half-solenoidal currents cover a large part of the dayside of the magnetosphere.

The two solenoidal currents join on the equatorial plane and flow together across the equatorial plane. This part of the current is called the *cross-tail*

128

current. Instabilities of the equatorial current play a crucial role during the expansion phase of auroral activities (Lui, 1991); see Section 6.3.

The cross-tail current is not connected to the ionosphere during a quiet time. It is during auroral substorms when it becomes connected to the ionosphere as we learn in Section 6.2 (a).

The above circuit can be shown in terms of a crude equivalent circuit: Figure 5.10 (left). The upper part is an inductive circuit (the magnetosphere) and is connected to the ionosphere (the lower part) by field-aligned currents, along which the aurora occurs. The orientation of the IMF acts like a *variable* brightness switch, since the orientation of the IMF varies between 0° (off) and 180°(on); more accurately, in the above power equation P, the factor $[\sin^4(\theta/2)]$ controls the intensity of the brightness of the aurora; the 'current sheet' represents the cross-tail current.

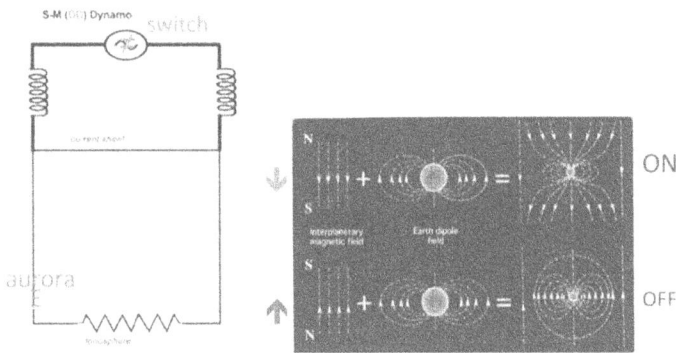

Figure 5.10 Left: Equivalent circuit of the magnetosphere-ionosphere coupling system in Figure 5.9. Right: The orientation of the interplanetary magnetic field acts as a variable brightness switch, the southward--"on" and northward--"off".

5.5 Monitoring ionospheric currents

Although electric current plays major roles in auroral physics and magnetospheric physics, there is no simple way to monitor continuously changes of electric currents in the upper part of the circuit in Figure 5.10

during auroral activities. The only *quantitative* way to monitor continuously the development of auroral substorms at present is to observe ionospheric currents, which are deduced from magnetic changes recorded by ground-based magnetometers as seen in the following.

We choose to monitor the total ionospheric current in the midnight-early morning sector, where the current is most intense. It may be noted that magnetic field observation at any point in the magnetosphere by a satellite cannot serve for monitoring auroral substorms.

(a) Setting up meridian chains of magnetic observatories

For this crucial purpose, a new close network of magnetic observatories had been set up to obtain accurately the distribution of the ionospheric currents; six meridian chains of magnetometers were set up during the International Magnetosphere Year in the 1970s and 1980s. Gordon Rostoker of the University of Alberta was instrumental in setting up two Canadian chains of magnetometers. Gordon and I met at Inuvik (the northernmost station in the arctic Canada) and cerebrated the completion of the Canadian chains. In Russia, A. N. Zaizev was in charge.

The Greenland chain was the responsibility of E. Friis-Christensen. I installed magnetometers at the Alaska meridian all-sky camera stations (Section 3.3). This magnetometer network study was supported partly by NSF. In Section 7.1 (c) and at the end of Chapter 7, it is pointed out that a similar method could be used in deducing the distribution of current on the photosphere of the sun.

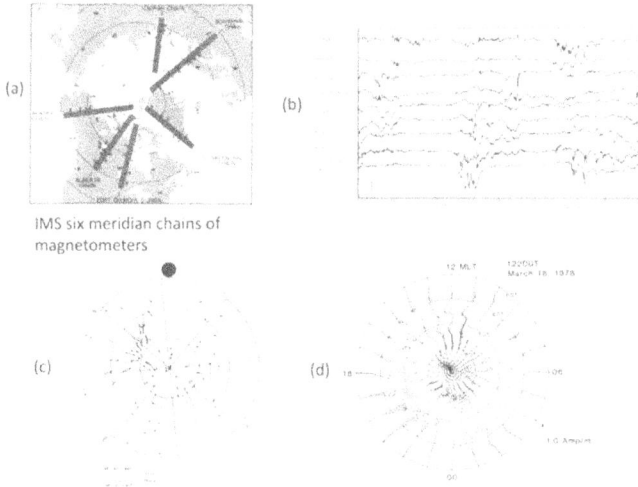

IMS six meridian chains of
magnetometers

Figure 5.11 (a)The international six meridian stations of magnetometers. (b) Example of observed magnetic records (H) from several chain stations. (c) The magnetic vector is determined at each station. (d) The special computer program converts the vector distribution to the distribution of ionospheric currents (Y. Kamide, 1982).

(b) Conversion of the observed magnetic vector to the current distribution

For this laborious project, Kamide et al. (1982) developed a special computer code to convert the observed magnetic vectors (horizontal component) into ionospheric current on a map. In this way, a large number of magnetograms from the six meridian chains were analyzed, with a time resolution of 5 minutes. With this extensive and laborious work, it became possible to learn how the ionospheric currents develop during auroral activities (auroral substorms) in great details.

The most intense westward current is called the *auroral electrojet* (Akasofu, et al.,1965); this term was coined by Chapman. It is a single cell current. The currents observed by this work are used extensively in this and next chapters. I thank Yosuke Kamide (my first post doc, from the Tokyo University) for his great effort.

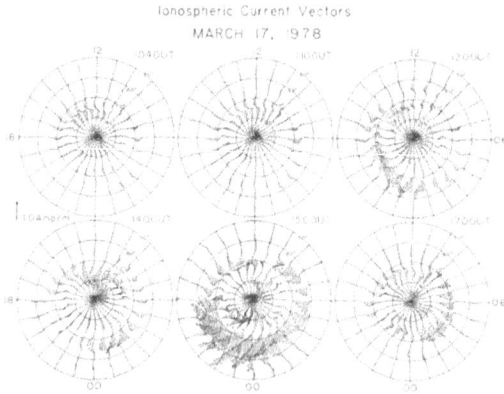

Figure 5.12 An example of changes of the electric currents in the ionosphere during auroral substorms (Kamide et al. 1982).

5.6 Separation of the DD and UL currents

The observed current in the ionosphere has two current systems together. The first one is the component of the primary M-I coupling system, which is directly connected the solar wind-magnetosphere dynamo. Its ionospheric part of this current is called *the directly driven (**DD**) current.*

Although details are given in the next chapter, two new current systems appear during the expansion phase. The first part is caused by the appearance of a new current system (Bostrom's meridional current) The second part is the disrupted cross-tail current, which flows into the ionosphere (Bostrom's azimuthal current system). These two currents together are called *the unloading (**UL**) current. Both currents constitute the auroral electrojet. Both are called* the secondary magnetosphere-ionosphere coupling current system (in short, the secondary M-I system).

The observed ionospheric current is (DD + UL) during the expansion phase, so that in order to study the expansion phase, it is crucial to learn on the UL current by separating it from the DD current; the AE index is a measure of (DD + UL).

Sun et al. (2000) succeeded in the separation of the DD and UL equivalent currents by a mathematical method (a sort of 2-D Fourier analysis) in terms of equivalent currents after a few years of efforts; it was fortunate that the UL current (the auroral electrojet) is a *single* cell current (Akasofu et al, 1965; Figure 5.13 [right]), while the DD current is a *two-cell* cell current, as seen Figure 5.13 (left).

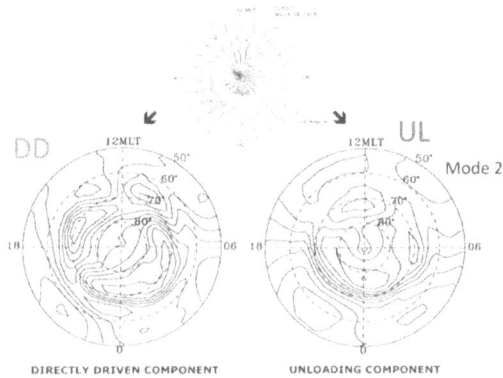

Figure 5.13 Separation of the ionospheric current (top) into the DD and UL currents (below) in terms of equivalent currents (Sun et al., 2000).

The success of the separation of the DD and UL currents can be tested by the fact that they can be identified by the established currents in the past. The DD current is a two-cell current (like Chapman's SD current, but distorted). This was confirmed by the superDARN (Super Dual Auroral Radar Network) radar study of Bristow and Jensen (2007). In Figure 5.14, the superDARN data are shown on the right. It may be compared with the DD current (middle). The superDARN network was led by R. A. Greenwald of the Johns Hopkins University (Greenwald et al., 1995). The plasma flow in the ionosphere is caused by the convective flow of magnetospheric and ionospheric plasmas (Axford and Hines, 1961).

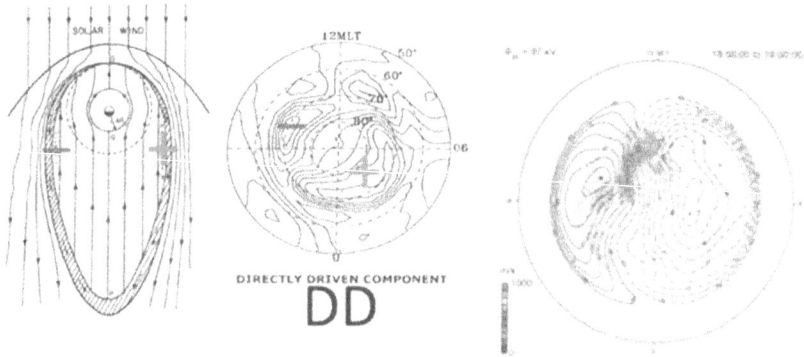

Figure 5.14 Left: The DD current is caused by the convective flow in the magnetosphere. Axford and Hines (1961) inferred the convective flow from Chapman's SD current. Middle: The potential pattern of the DD current, showing a two-cell pattern. Right: The ionospheric convection pattern obtained by the superDARN network (Bristow and Jensen, 2007).

On the other hand, the UL current is known to be a single cell current. (Akasofu et al., 1965). In Figure 5.15, it can be seen that the auroral electrojet is mostly the UL current.

In order to study the expansion phase, it is crucial to know characteristics of the UL current. As we will see in Section 6.3 (Figure 6.4 b), the UL current is found to be impulsive (manifested by the auroral activity during the expansion phase). Without this time variation of the UL current, it is difficult to study quantitatively the explosive expansion phase of auroral substorms (why and how the UL current is impulsive).

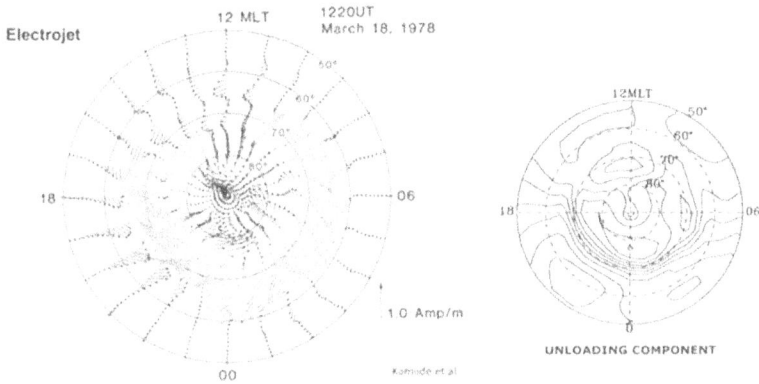

Figure 5.15 Left: An example of the observed current distribution (Kamide). Right: The UL equivalent current, indicating that the main westward current on the left is the UL current.

Figure 5.16 Left: Sun Wei. He was very talented person, artist and dancer. He and I made a Chinese poem about the four seasons of Alaska.

I owe Sun Wei for his great effort of analyzing magnetic records in my substorm research, particularly for the success in the separation of the DD and UL components.

5.7 Variations of the DD and UL currents

Now, we learn how the DD current and the UL current vary during the growth, expansion and recovery phases.

135

During the growth phase, the DD current does not grow, although we expect from the fact that the power becomes above 10^{11} W. This is consistent with the fact that there is no specific auroral activity during the growth phase, namely no clear dissipation of the power. These are crucial facts in understanding the significance of the growth phase. During the expansion phase, the UL current develops suddenly and subsides quickly in about one hour, namely an impulsive occurrence. We see both DD and UL currents as a function of time in Section 6.2, together with the power as a function of time.

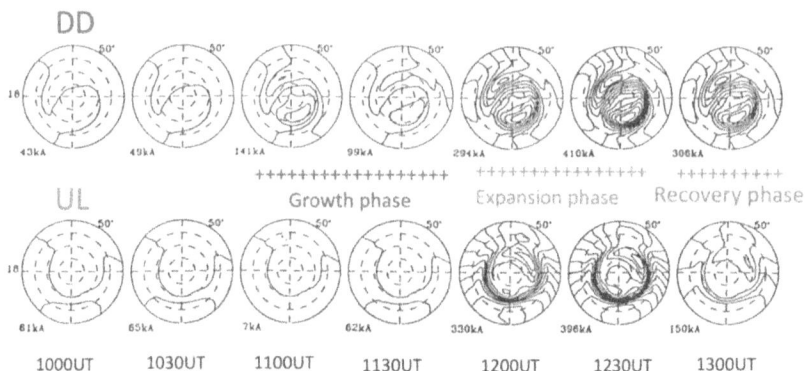

Figure 5.17 Development of the ionospheric currents. The DD and UL components are separately shown.

In Section 6.3, it will be seen that this result plays also the vital role in understanding the explosive feature of the expansion phase.

5.8 Joule heat production

One of the most important quantities in studying auroral activities is the dissipation rate (W or erg/s) and the total dissipation (J or erg), namely how much power an auroral substorm consumes. So far, the Joule heat dissipation is the only way to know these two quantities semi-quantitatively as a function of time in a study of auroral substorms.

This study can be done only (at present) on the basis of the ionospheric electric current. Fortunately, the Joule heating is the largest ones, since the particle dissipation is only 1/10 of the Joule heat dissipation (Ahn et al. 1983). Ahn was one of my graduate students, studying this problem on the basis of the current determined by Kamide et al. (1982) and the Chatanika Incoherent Scatter radar data. An example of his results is shown in Figure 5.18.

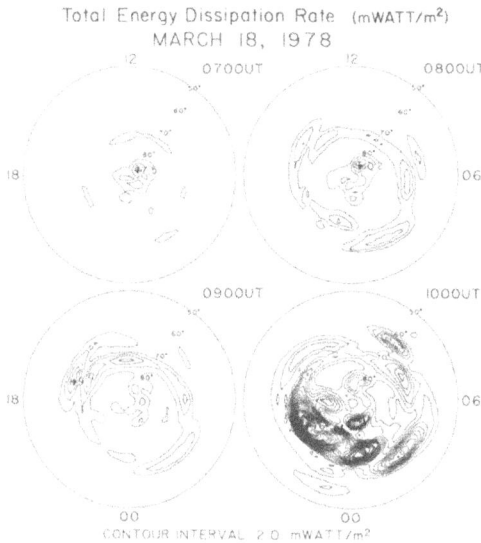

Figure 5.18 The development of the Joule heating in the ionosphere during a typical substorm (Ahn et al.,1983). Note that 1 mw/m^2 = 1 erg/cm^{2}s. This corresponds to the current distribution in Figure 5.12.

Figure 5.19 With B.-H. Ahn in Korea. He made an important study of the Joule dissipation.

In this way, we can monitor how the Joule heating rate develops in the ionosphere during a typical auroral activities (auroral substorms). Its integrated value over the entire polar region is the total amount of the Joule heat production during an auroral substorm. In this study, the Chatanika incoherent radar observation led by Peter M. Banks was also very useful (Brekke et al.,1973).

Figure 5.20 With Y. Kamide (right) and Tony Lui (middle) at a volcano in Hokkaido. Both worked closely with me in analyzing the magnetic and auroral data.

My motive

When I began to learn MHD, I was impressed by its ability to explain many astrophysical phenomena. However, when I met Hannes Alfven in 1967, he told me that all problems which can be understood had already solved. All unsolved problems cannot be solved by MHD. I was surprised by what he said, because he established MHD, and I just began to study it.

He suggested that I should take the electric current approach.

However, it is not possible to measure *directly* electric current in space physics, either ground-based or satellite-based instruments. Satellites can measure magnetic field only at a single point. Fortunately, there had traditionally been an extensive study of ionospheric current which was deduced by ground-based magnetometers. Thus, one of the first steps was to establish a good network of magnetometers, six meridian chains of magnetosphere. The second step was to deduce the distribution of electric current in the ionosphere.

I had been greatly interested in electric current which flows in the ionosphere. Chapman's 2-D (SD or DS on a spherical surface) ionospheric current was well-established fact at that time. But, I found a paper by A. P. Nikolsky (1947, see Episode in this chapter), in which he showed that current system in the polar region differs from Chapman's one (Chapman and Bartels, 1940). However, most people did not believe Nikolsky's result. In our auroral study, Akasofu et al. (1965) found that the auroral electric current flows along the auroral oval (Nikolsky's result), not along the auroral zone (as the SD current shows).

In addition, Chapman's 2-D current was criticized by researchers who believed Birkeland's 3-D current system (they called it "the Birkeland current"). Since the definition of field-aligned current is the current between the magnetosphere and the ionosphere, I wanted to clarify the controversy. I found that Chapman stated earlier that his current was "equivalent" current on a spherical surface by stating that ground-based magnetic observations alone cannot determine where the current is located; his study was made before the ionosphere was well known. After the ionosphere was discovered, people identified Chapman's spherical surface as the ionosphere.

For these reasons, I was reasonably familiar with ionospheric currents and did not hesitate to take up the electric current approach. Another simple reason to take up the electric current approach is that the aurora is an electrical discharge phenomenon, so that it is natural to consider a dynamo as the power generator.

In the 1960s and 1970s (and even now), most researchers were interested in satellite data; ground-based magnetograms were used to determine only 'substorm onset', not interested in the current system. They relied on the storm indices like the AE or Dst index.

Episodes

(1) Jack Townshend

Jack Townshend was the chief of the College Observatory. Jack played an important role in our research by providing us high quality magnetograms. He published College monthly magnetograms, which were widely used by space physics researchers in the world as "College magnetograms". I frequently visited his observatory, which is located near the Geophysical Institute of the University of Alaska Fairbanks. He presented me one of their variometers when he replaced them with new electronic magnetometers.

(2) A. P. Nikolsky

A. P. Nikolsky (1947) found magnetic disturbances tend to occur along peculiar spiral lines in the polar region (Figure [b] below). By examining details of polar magnetic records, I found his work was correct; magnetic disturbances occur along the *auroral oval*, not along the *auroral zone*.

I gave a talk in Moscow about it. After my talk, Nikolsky came to the stage and gave me a Russian bear hug, saying I was very young like his son. His work had been almost disregarded at that time.

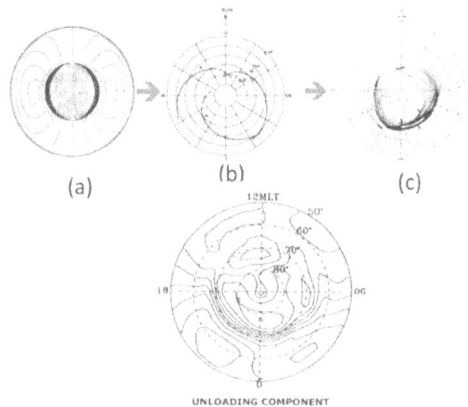

(a) Chapman's famous SD ionospheric current. (b) Nikolski thought that magnetic disturbances occur along spiral curves. (c) Akasofu, Chapman and Meng (1965) found that during auroral activities, the current system has a single cell along Nikolsky's spiral curves. (d) The UL single cell current is more accurately determined (Figure 5.15).

(3) Polar magnetic changes during the extremely quiet days

The magnetic changes during extremely quiet days (Kp = 0000 0000) is obtained from the magnetic stations all over the world (Kawasaki and Akasofu 1972).

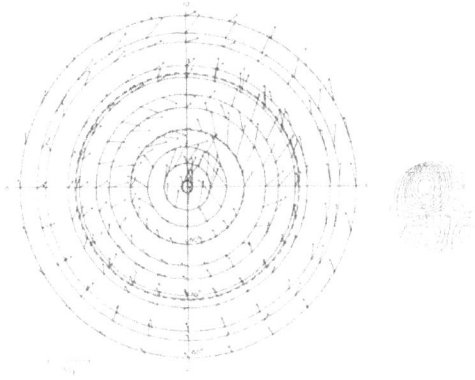

Magnetic changes during extremely quiet days (Kp = 0000 0000); Kawasaki and Akasofu (1967). One can see the quiet day daily variation Sq. One can see also magnetic fields in the highest latitude region. The attached the Sq current (Chapman and Bartels, 1950).

(4) Effects of the aurora

We studied possible problems caused by the aurora. One of them is to examine the corrosion of the trans-Alaska pipeline by the aurora-induced electric current. The auroral current induces electric currents on the ground, and its part is channeled into the 800-mile-long pipe, which corrodes it. The pipeline company placed a magnesium alloy along the pipe to prevent the corrosion.

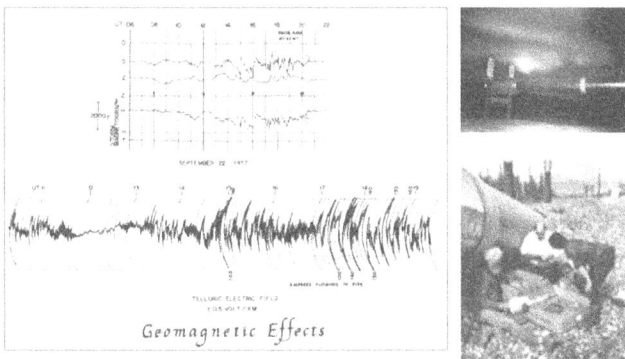

Trans-Alaska Oil Pipeline and the aurora (with Gene Wescott and Bob Merritt). The figure shows also the College (insensitive) magnetogram and the current in the pipeline during an intense geomagnetic storm.

Another problem is aurora-induced currents in power transmission lines. Fortunately, the electric company allowed us to install a set of instruments for this purpose. We found that the induced currents in the circuit breaker are impulsive.

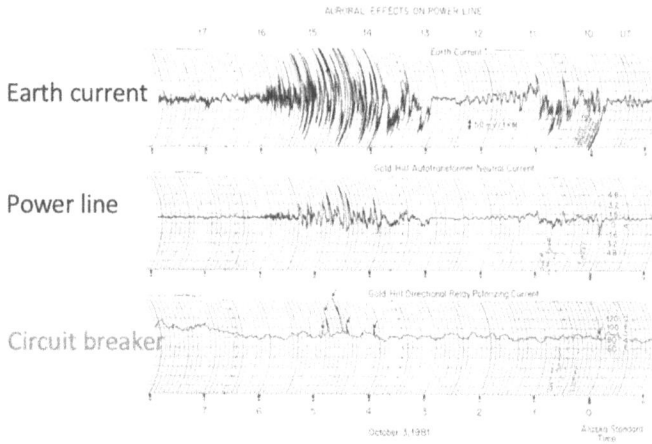

Earth current

Power line

Circuit breaker

Power line and the aurora. From the top, the earth current record, the current in the transmission line. The breaker circuit shows auroral effects as pulses (Aspnes et al., 1981).

References

Ahn, B.-H., Akasofu, S.-I. and Kamide, Y., 1983, The Joule heat production rate and the particle injection rate as a function of the geomagnetic indices AE and AL, J. Geophys. Res. **88**, 6275.

Akasofu, S.-I., Chapman, S. and Meng, C.-I., 1965, The polar electrojet, J. Atmosph. Terr., Phys. **27**,1301.

Akasofu, S-. I. 1981, Energy coupling between the solar wind and the magnetosphere, Space Sci. Rev. **28**,121.

Akasofu, S.-I.,1985, The polar caps. National Institute of Polar Research Special issue, No. 38, 1.

Aspnes, J. D., Merritt, R. P. and Akasofu, S.-I., 1981, Effects of geomagnetically induced current on electric power systems, The North. Eng., **13**, 34

Axford, W. I. and Hines, C. O., 1961, A unifying theory of high-latitude geophysical phenomena and geomagnetic storms, Can. J. Phys., **39**, 1433.

Brekke, A., Doupnik, J. R. and Banks, P. M., 1974, Incoherent scatter measurements of E region conductivities and currents in the auroral zone, J. Geophys. Res.,**79**, 3773.

Bristow, W. A. and Jensen, P., 2007, A superposed epoch study of superDarn convection observations during substorms, J. Geophys. Res., **69**, 112, A06232, doi:10. 1029/2006JA012049.

Greenwald, R. A., et al., 1995, DAN/SuperDARN: A global view of high-latitude convection, Space Sci. Rev., **71**, 763.

Kamide, Y. et al., 1982, Global distribution of ionospheric and field-aligned currents during substorms as determined from six IMS meridian chains of magnetometers: Initial results, J. Geophys. Res., **87**, 8,228.

Kawasaki, K. and Akasofu, S.-I., 1967, Polar solar daily geomagnetic variations on exceptionally quiet days, J. Geophys. Res., **72**, 5,363.

Nikolsky, A. P., 1947, Dual laws of the course of magnetic disturbances and the nature of mean regular variations, Terr. Mag. Atmos. Elect.**, 52,**147.

Olson, W. P., 1984, Introduction to the topology of the magnetospheric current systems, 49 in Magnetospheric Currents, ed. By T. A. Potemra, AGU Monograph vol. **28**.

Perreault, P. and Akasofu, S.-I., 1978, A study of geomagnetic storms, *Geophys. J. Roy. Astron. Soc., 54,* 547.

Sun, W., Xu, S.-Y., and Akasofu, S.-I., 2000, Mathematical separation of directly driven and unloading components in the ionospheric equivalent current during substorms, J. Geophys. Res., **103**, 11695.

Vasyliunas, V.M., Kan, J. R., G.L. Siscoe, G. L. and S.-I. Akasofu, S.-I, 1982, Scaling relations governing magnetospheric energy transfer, *Planet. Space Sci., 30*, 359.

Chapter 6 Auroral substorms: Searching for the cause of the explosive nature

———————— ▷ ▶ ▣ ⊹ ⊰ ◀ ◁ ————————

During several decades, a large number of ground-based and satellite-based observations have become available. Many of them have been theoretically interpreted. Therefore, one of the next tasks in substorm research is to *try to synthesize them or sequence them* in order to understand how auroral substorms can occur and develop, in particular the explosive expansion phase. Thus, the purpose in this chapter is to establish a morphological theory of auroral substorms in this way.

In this chapter, I attempt to synthesize these observations and physical processes described in the earlier and present chapters by considering the magnetosphere as an *electrical circuit system and by following energy flow in terms* of the *electric current approach* (Alfven, 1968,1981). In general, synthesis efforts would not succeed without a specific basic principle in mind.

The first step in this attempt is to find the power supply. As we discussed in Section 5.2, it is necessary to consider that the solar wind-magnetosphere interaction constitutes a dynamo, the *auroral dynamo*.

Thus, we see in this chapter that the power generated by the dynamo flows first toward the inner magnetosphere and is accumulated in the inner magnetosphere, causing *inflation of the inner magnetosphere.* In terms of electric current, the cross-tail current is enhanced particularly around 6 Re. This is based on the fact that there is no auroral and magnetic activities during the growth phase, so that the power is not dissipated during the growth phase, and thus is accumulated.

In the second step, we see that the magnetosphere becomes unstable, when the accumulate energy exceeds 10^{16} J as a result of instability of the cross-tail current. This amount of 10^{16} J seems to be the upper limit for the inner magnetosphere (6 Re) can hold the accumulated energy.

The instability causes a sudden reduction of the cross-tail current, causing *deflation of the inner magnetosphere*. The deflation causes charge separation, which results in an earthward electric field, which in turn results in a new current system (Bostrom's meridional current system) with field-aligned current, which in turn ionizes the ionosphere, causing the sudden brightening of the arc. The increased ionization makes the ionosphere conductive enough for the disrupted cross-tail current to flow into the ionosphere (Bostrom's azimuthal current system), dissipating the accumulated energy as the Joule heat. This whole sequence of processes causes various aspects of auroral substorms.

The purpose of this chapter is to explain this whole sequence with observed data.

On the other hand, the magnetic field line approach considers that the IMF field lines link with magnetic field lines at the front side of the magnetosphere, and the linked field lines are carried towards the magnetotail by the solar wind. The energy is accumulated as magnetic energy in the anti-parallel magnetic configuration in the magnetotail.

In this chapter, we can see that it is not possible to follow quantitively the development of auroral substorms as a function of time without monitoring electric current. The electric current approach is so far the only way to know how the expansion phase develops and how much the energy is dissipated as a function of time. The magnetic field line approach has to rely on the unknown rate of magnetic reconnection.

Further, one of key points in our approach is the formation of the double layer, a U-shaped potential structure (the electric field along magnetic field lines), which is essential in accelerating current-carrying-electrons in the field-aligned current to 10 KeV from 300 eV in the magnetosphere. Since we deal with field-aligned current in the electrical circuit, the double layer can naturally be introduced in the circuit. Without it, the whole circuit is open, and there would not be the aurora.

The magnetic field line approach alone cannot have the double layer, because an electric field along magnetic field lines is not allowed in MHD. Without the double layer, there is no simple way to generate 10 KeV electrons in the magnetosphere.

6.1 Electric current circuit

A study of auroral substorms in terms of an electrical circuit is, however, *practically a new attempt*, although Alfven (1968) considered it first conceptually in as early as 1968. The reason for 'new' is because the magnetic field line approach has almost exclusively has been adopted until today.

Actually, I was greatly surprised to find Alfven's statement in his paper in discussing the magnetospheric circuit: "Akasofu is the only one in the list who has understood the [circuit] value of this cosmic physics"; (Alfven, 1986, p.796).

Figure 6.1 With Hannes Alfven (GI).

Figure 6.2 represents the current circuit in its simplest circuit format; its more realistic version was shown in Figure 5.9. The upper part of Figure 6.2 represents the magnetosphere; the sheet current represents the cross-tail current. The magnetosphere is connected to the ionosphere by field-aligned currents. The aurora is seen in the ionosphere along field-aligned current.

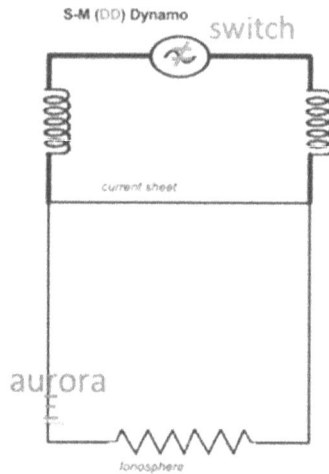

Figure 6.2 Magnetosphere-ionosphere discharge circuit (equivalent). The aurora is the visible part of the circuit at the boundary between the magnetosphere and the ionosphere. The aurora is the phenomenon which occurs as the only visible phenomenon in the ionosphere within the whole circuit. The 'current sheet' represents the cross-tail current.

In this chapter, we consider auroral substorms as the sequence of processes in the magnetosphere-ionosphere coupling system by considering the above electrical circuit. Thus, our synthesis is guided by this basic equivalent circuit. In the following, this simple circuit is shown for the three phases of substorms, the growth, expansion and recovery phases.

One of the reasons for adopting the circuit diagram is that one can keep always in mind the whole electrical system of the magnetosphere, even when one is considering one part in the sequence.

As we discuss in the following, the accumulation of power occurs in the inductive part of the circuit during the growth phase; it is suggested that this occurs because of a high resistivity of the ionosphere (or a low conductivity) during the growth phase.

Then, during the expansion phase, the development of the secondary magnetosphere-ionosphere coupling system (red circuit in the middle circuit) occurs by the transfer of the accumulated energy. It is this circuit which generates the UL current (Section 5.6), starting the expansion phase. Then, the increased conductivity of the ionosphere allows the disrupted cross-tail current to flow into the ionosphere. Note that the cross-tail current can flow into the ionosphere, because the high ionospheric resistivity is reduced during the expansion phase.

We refer to Figure 6.3 later in discussing each phase of substorms.

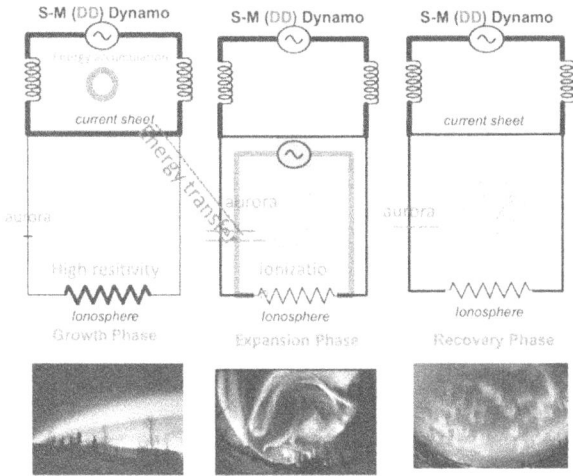

Figure 6.3 Electrical circuit for the three phases of auroral substorms, together with a typical auroral activity in each phase. Note the changes of the ionosphere and current sheet.

6.2 Inflation of the magnetosphere: The growth phase

After learning how the power is generated in the previous chapter, the next question is where the power is transmitted or flows.

When the solar wind-magnetosphere dynamo power ($\varepsilon = P$) begins to increase above 10^{11} W (Section 5.7, Figure 5.17), we observed that the DD current does not grow during the growth phase. In Figure 6.4, we show again the currents DD and UL (Figure 5.17), but also together with the power ε (= P) as a function of time.

Figure 6.4 is crucial in that it is the only way to learn *observationally* how auroral substorms develop quantitatively in time (Akasofu.2017), particularly because we can see the DD and UL currents separately (not [DD +UL], like the AE index).

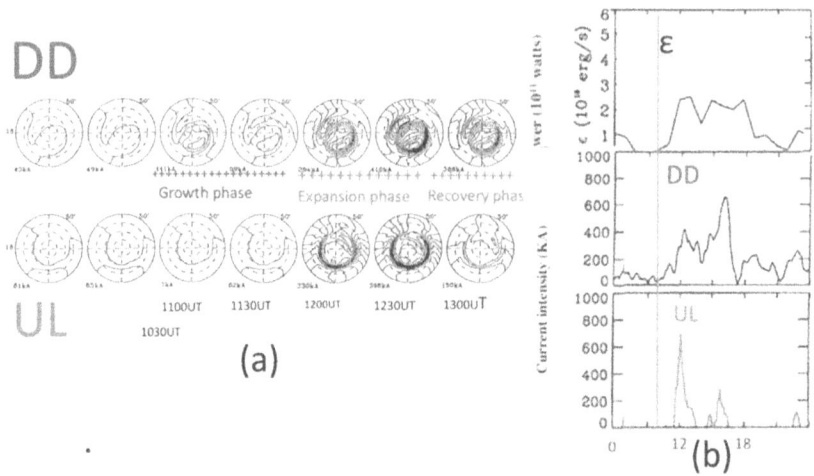

(a)

(b)

Figure 6.4 (a)Development of ionospheric currents, both the DD and UL components during a typical auroral substorm. (b) The time variation of the power ε (= P) together with the time variation of the DD and UL currents.

First of all, it can be seen again that the DD current did not seem to grow well with the power during the growth phase period (11:00-11:30 UT) until the UL current suddenly occurred (expansion phase onset).

The fact that the DD current cannot flow well during the growth phase suggests that the ionosphere cannot let the DD current to flow, because of its low conductivity (or the field-aligned current is too weak to develop the double layer; Section 6.4); this situation is illustrated in Figure 6.3 (growth phase circuit), in which the ionosphere has a high resistivity, and the power is accumulated in the upper inductive circuit.

This is supported by the fact that there occurs no specific auroral activity before the sudden brightening of the arc, expansion phase onset, as shown below. Further, as we can see later, there is no new electron flux or no change during the growth phase (Figure 6.6); thus, although there is slight increase of the DD current in this case, but its origin is uncertain.

In fact, in Figure 6.5a, the arc was not present from 10:00 to 11:09 UT, about one hour before the sudden appearance of an arc, namely during the growth phase. Although there was some auroral activity in the poleward sky within the auroral oval and such a poleward auroral activity is common; it is not a specific and systematic activity and seems to be unrelated to the initial brightening of arcs; actually, in this case, the northward arc in the auroral oval brightened when the equatorward arc appeared and brightened.

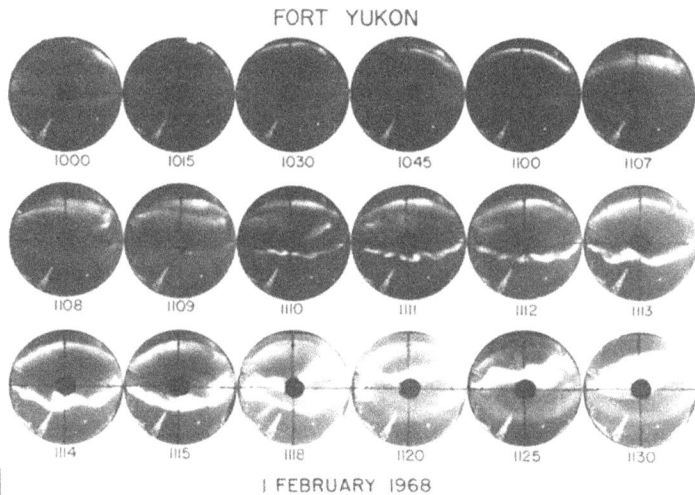

Figure 6.5a An example of a sudden appearance of an arc and its immediate poleward advance. Note that there was no specific auroral activity prior to the sudden brightening (University of Alaska Fairbanks).

An example of satellite images in Figure 6.5b shows clearly no specific auroral activity in the poleward of the polar region before about one hour before the onset.

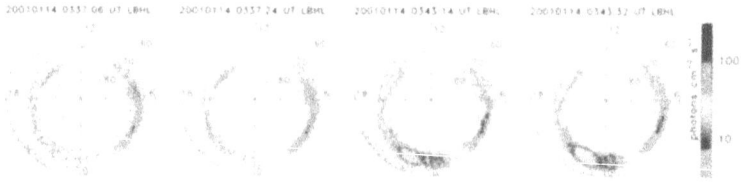

Figure 6.5b The same as Figure 4.12. No specific auroral activity before the onset.

Thus, the next question is where does the power go?

Actually, the magnetosphere has no choice, but to accumulate the incoming power in the inductive part of the primary M-I system.

Since the dynamo process occurs from the front part of the magnetosphere and the flow of the dynamo power follows along the Poynting flux ($E \times B$), a large part of the dynamo-generated power flows toward *the inner magnetosphere* as Figure 6.6 indicates (blue arrows toward the equator of a distance of 6 Re) for medium intensity substorms or the IMF Bz intensity of about -5 nT (Akasofu, 2017a).

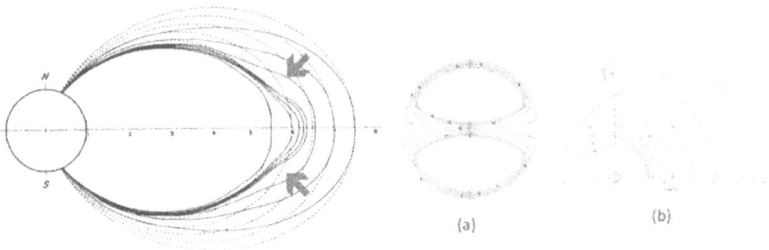

Figure 6.6 Left: Direction of the Poynting flux ($E \times B$) generated by the dynamo. It is directed mainly toward the inner magnetosphere. The deformation (inflation) is the result of simulation of the accumulated energy of 10^{16}J. Right: Same as Figure 5.9. (a)The two half-solenoidal currents along the cross-section of the magnetosphere. (b) Joint current on the equatorial current, the cross-tail current (Olson, 1984).

The growth phase is thus the power accumulation period, namely the period between the power becomes above 10^{11} W and the onset of the sudden growth of the UL current or the sudden brightening of an auroral arc. The power 10^{11} W is more accurate than IMF Bz becomes negative (see Figure 5.8).

The result of this accumulation can also be understood in terms of an increase of the equatorial cross-tail current as the dynamo power increases; the dynamo current at a distance of 6 Re is most intense (right part of Figure 6.6 or Figure 5.9c). Thus, the magnetic energy is accumulated mainly in the inner magnetosphere, causing the *inflation* of the inner magnetosphere. The fact that the outer boundary of the outer radiation belt and the aural oval coincide is also another important proof of the accumulation of energy in the inner magnetosphere, not much in the magnetotail (Section 3.3).

Thus, one of the major differences between the electric current approach and the magnetic field line approach is the location where the accumulated energy is stored, either in the inner magnetosphere or in the magnetotail.

The inflation in the inner magnetosphere can be observed by geosynchronous satellites at a distance of 6 Re as a *decrease* (50 nT) of the magnetic field (the blue arrows in Figure 6.7); Deforest and McIlwain (1971). This decrease is caused by diamagnetism of the cross-tail current (Section 2.3 [c]).

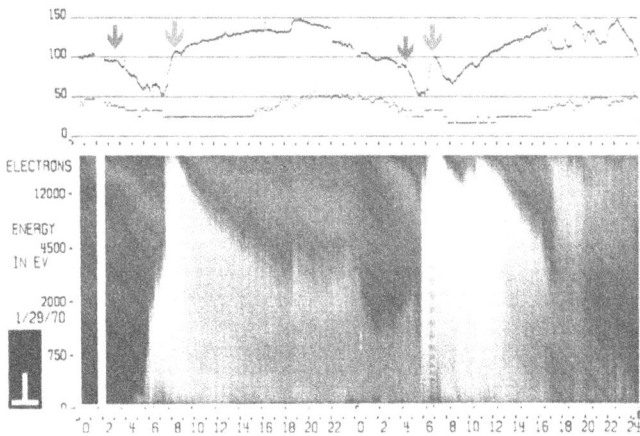

Figure 6.7 Both magnetic field and electron observations at 6 Re by the ATS satellite. Two substorms are observed in this record. Note that the inflations (a decrease of the intensity of the magnetic field, blue arrows) occurred about one hour before the expansion phase onset (shown by a sudden appearance of intense electron flux, the red arrows); the period between the magnetic decrease and the sharp increase of the electron flux is the growth phase (Deforest and McIlwain, 1971). The red arrow will be discussed in the following.

Note also in Figure 6.7, as mentioned earlier, there is no new electron flux or its variations during the inflation (when the magnetic field is decreasing); this is consistent with the fact that there is no specific magnetic and auroral activity during the growth phase, proving that there is no new ionization/dissipation in the ionosphere during the growth phase.

The amount of energy accumulated in the inner magnetosphere (10^{16} J) is estimated from the energy spent during the expansion phase based on the ground-based observation of the Joule heat production; other reasons are given in the following. *Thus, this amount is the limit of energy, which the inner magnetosphere can accumulate at about the distance of 6 Re.* In other words, this is why auroral substorms of medium intensity occur in the way they occur. If the accumulation can occur at a shorter distance, substorms may be more intense. This is because the magnetic dipole intensity is stronger, allowing a larger amount of energy to maintain. Recall also that the auroral oval expands to lower latitudes (Section 3.4)

It so happened that the simulation is also close to the limit of its method because the magnetic field becomes too weak for the guiding center approximation at the center of the inflation. The inflation of magnetosphere by the simulation is shown in Figures 6.6 and 6.8 corresponds to the accumulated energy is about 10^{16} J.

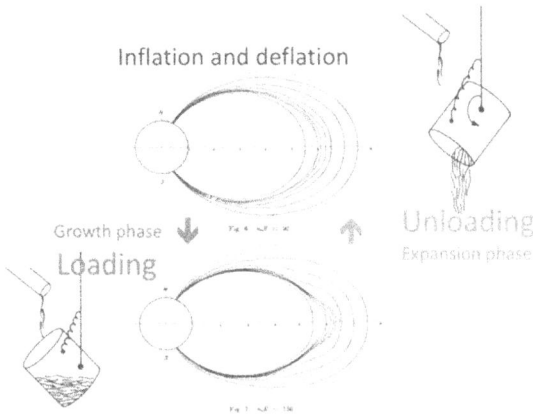

Figure 6.8 The inflation/deflation of the magnetosphere by the loading/unloading. The simulation of the inflation of the magnetosphere with the accumulated energy of about 10^{16} J.

There are observationally four crucial points in understanding theoretically the cause of the expansion phase.

(1) Location of the sudden brightening

The location of the energy accumulation, in addition to the direction of the Poynting flux, is closely related to the location of the first indication of the expansion onset (a sudden brightening of an auroral arc). It is located near *the southern (equatorward)* boundary of the auroral oval; Figure 4.9 shows an observational confirmation of this fact.

The location of the initial brightening arc is, on the average, about 65° in latitude for medium intensity substorms, which is connected to the equatorial plane at about 6 Re by the magnetic field lines. This location corresponds to the outer boundary of the outer radiation belt — as Van Allen and I found — not the magnetotail; Sections 3.3 and 4.3.

As mentioned earlier in this section, there is no indication that auroral arcs located northward of the sudden brightening arcs in the oval show any systematic change, so that the sudden brightening of arc is triggered by an internal cause, not by the magnetotail.

(2) Impulsive feature

Another important fact in Figure 6.4 is an *impulsive development of the UL current and the expansion phase*. The impulsive development of the UL current represents and corresponds to the explosive feature of auroral displays during the expansion phase in terms of the electric current.

The impulsive development of the UL current in Figure 6.4 (right) can tell quantitatively how the expansion phase develops; the peak period is rather short, it is about one hour.

Thus, we have to explain why the expansion phase is impulsive, in addition to the fact that the onset is not triggered by an increase of the power.

(3) Occurrence of the UL current

It is well-established that the expansion phase occurs about 40-100 minutes after the power increases above 10^{11} W, namely the period of the growth phase.

Another crucial point in the UL current is that *it appears during an early phase of (40-100 minutes after power begins to increase above 10^{11} W) substorms, although substorms last, in general, for 2-3 hours*; Figure 6.4). The UL current *does not appear afterward unless a significant power increase occurs* (Akasofu, 2017a,b).

(4) Amount of energy released

The amount of energy accumulated in the inner magnetosphere (10^{16} J) is estimated from the energy spent during the expansion phase based on the ground-based observation of the Joule heat production (Section 5.8). As mentioned earlier, this amount is the limit of energy, which can be accumulated at about the distance of 6 Re in the inner magnetosphere.

Thus, it is quite likely that the accumulated energy during the growth phase is almost completely released during the expansion phase. The released energy is about 10^{16} J as mentioned earlier. So far, the ground-based study is the only way to determine the amount of about 10^{16} J, which is important to determine the amount is of energy, which the inner magnetosphere can hold for medium intensity substorms (AE $\approx$ 500 nT), which is also closely related to the instability of the cross-tail current as we discuss in the following.

In summary:

We attempted synthesize various observations, their interpretations and theoretical considerations in sequence during the growth phase. It may be worthwhile to review our attempt during the growth phase in the figure below.

The growth phase is found to be the energy accumulation period. The expansion period is the energy release period. Figure 6.9 shows the above synthesis.

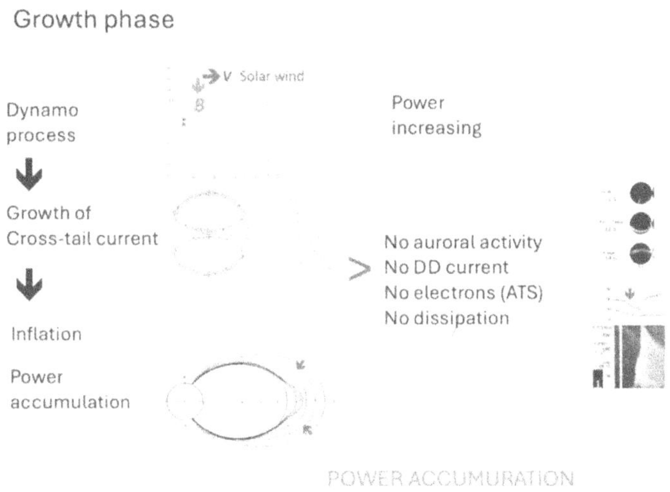

Figure 6.9 Synthesis of the growth phase phenomena.

6.3 Deflation of the magnetosphere: The expansion phase

When the accumulated amount exceeds about 10^{16} J during the growth phase, the cross-tail current seems to develop a specific instability.

(a) Disruption of the equatorial current

The limitation of the accumulated amount of energy may be closely related to the limit of the intensity of the cross-tail current at 6 Re, where it is most intense (Figure 6.6 or Figure 5.9 (right [c]). The nature of the current instability associated with this magnetospheric instability has been studied by Lui (1991, 2011, 2020). The current instability reduces or disrupts the cross-tail current. As a result, the disrupted current is directed to the ionosphere, where its energy is dissipated as the Joule heat there. In the following how this happens is examined. This part is a joint work with Tony Lui and Lou Lee (Akasofu et al., 2025).

The way for the disrupted current to flow into the ionosphere is indicated in the current system shown in the circuit (*cefd*) in Figure 6.10; the cross-tail current flows between *a* and *b* during a quiet time, and it is disrupted between *c* and *d* during the expansion phase. It is the circuit (*cefd*) is the *discharge circuit*.

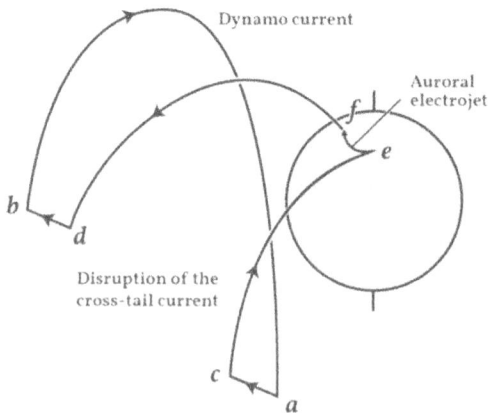

Figure 6.10 Schematic illustration of the disruption of the cross-tail current and the resulting the field-aligned current and the auroral electrojet. The line connecting *a* and *b* is the dynamo current (within the dynamo, the current flows from the negative terminal to the positive terminal). (2) The equatorial current is disrupted between *c* and *d*. (3) The disrupted current flows along c-e-f-d. The currents c-e and f-d are field-aligned currents. (4) the current e-f is the auroral electrojet, the UL current.

158

The reduced or disrupted cross-tail current between *c* and *d* is expected to flow into the ionosphere as the UL current in the ionosphere where the accumulated energy is dissipated as the Joule heat (Section 5.8).

However, the disrupted cross-tail current cannot flow into the ionosphere during a quiet period, because the quiet ionosphere (*ef*) is not conductive enough (or the double layer is too weak to generate 10 KeV electrons, see Section 6.4) as we saw during the growth phase.

Although such a current system is often used by many, such a current system alone cannot explain many auroral phenomena during the expansion phase of auroral substorms, such as the initial brightening of the arc *in a wide longitude range*; in Figure 6.10, the only place the aurora can occur is the point *f*, where there is an upward field-aligned current.

In order to allow the ionosphere to receive the disrupted cross-tail current between *c* and *d* or to establish the current system in Figure 6.10, there occur two stages of process.

Stage I: Bostrom's Meridional current system

The magnetosphere develops a new current system. It is the red circuit in the middle of Figure 6.3. In order to understand this process, we have to rely on a set of crucial observations in the following.

The observed sequence of events associated with the current reduction is shown by Lui (2011); Figure 6.11 (a).

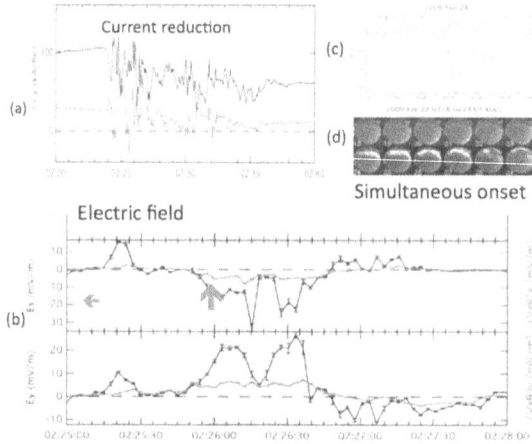

Figure 6.11 Satellite observation at 8.1 Re (Lui, 2011). (a) The reduction of the cross-tail current (black). (b) The growth of an earthward electric field (red arrow), together with the breakdown of the frozen-in field condition (red line). (c) The simultaneous occurrence of the electrojet (magnetogram) and (d) all-sky photographs.

Fortunately, this set of observations contains many hints of how the magnetosphere increases the conductivity of the ionosphere or how the sequence of following processes occurs for the expansion phase.

The crucial observation is the occurrence of an *earthward electric field (b) associated with the sudden reduction of the cross-tail current (a) in Figure 6.11.*

One way of understanding the occurrence of the earthward electric field is that the disruption of the cross-tail current causes the *deflation* (after the inflation during the growth phase). During the deflation, charge separation is expected to occur, resulting in the earthward electric field, because electrons are tightly bound to the contracting field lines (moving toward the earth), while protons may not (Lui and Kamide, 2003); Figure 6.12.

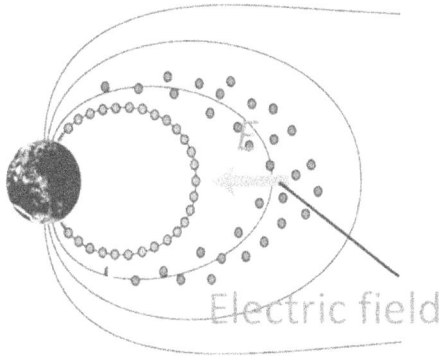

Figure 6.12 Electron are moving toward the earth with the contracting magnetic field lines with the contracting field lines, while protons are almost gyro-free (because a very low field intensity in the region of the inflation), so that they tend to remain where they are. Thus, the charge separation occurs during the deflation.

This is because the magnetic field intensity is very weak at the location of the inflation (diamagnetism) for the protons to perform the gyration properly; this is another reason why substorms originate at about 6 Re. Thus, the "frozen-in" field concept (MHD) breaks down at this crucial time at this location. This electric field is set up on the equatorial plane over a wide range in longitude as the initial brightening arc demonstrates.

Another way of understanding the occurrence of the earthward electric field is that the deflation causes an increase of the magnetic field at 6 Re (Figure 6.7, red arrows). The increase is called the *dipolarization*. The occurrence of this earthward electric field can also be understood in terms of the dipolarization (the red arrows in Figure 6.7), $E = [-(\partial Bz/\partial t)\int \partial y] \approx 5\text{-}50 \text{ mV/m}$.

This earthward-oriented electric field generates a new current system, the secondary magnetosphere-ionosphere coupling system (in short, the secondary M-I system), the red circuit in Figure 6.3 (The primary M-I system was mentioned in Section 5.4). We identify this new current system, which was originally proposed by Bostrom (1946).

The electrons which are shifted earthward by contracting field lines are discharged along the magnetic field lines, generating the field-aligned current in a *sheet* form as shown in Figure 6.13. It is this *sheet field-aligned current*, which causes the initial brightening arc at substorm onset over a wide longitude range as the shifted electrons interact with the ionosphere. *This is the first discharge process.* This situation is also illustrated in Figure 6.2 (expansion phase), in which the new current system is shown in red. It is this current system, which is responsible for the initial brightening of an arc (Figure 6.13).

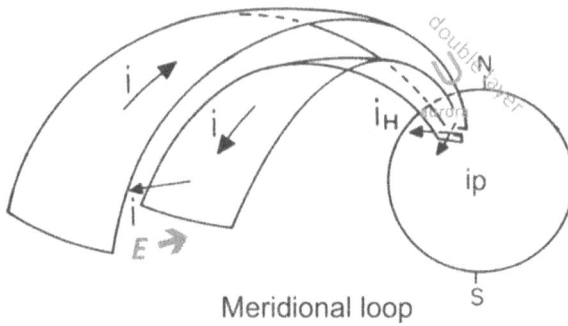

Meridional loop

Figure 6.13 Meridional current system produced by the charge separation, originally proposed by Bostrom (1964). Note that the field-aligned current is an extensive (in longitude) sheet current. This current system is generated by the earthward-directed E during the deflation. The double layer is also shown in red.

The extensive *sheet form of field-aligned current can explain a curtain-like form of auroral arc* along a great distance in longitude. Further, field-aligned current can develop a double layer that can accelerate current-carrying electrons more than 10 KeV (Section 6.4), so that they can ionize the lower ionosphere over a wide longitudinal range at once, explaining substorm onset. We apply a similar consideration in explaining two-ribbon structure of solar flares (Section 7.1 [e]).

Thie location of the initial brightening arc is the latitude of about 65°, which is about the *southern boundary of the auroral oval.* Thus, the disrupted cross-tail current flows into the ionosphere along the magnetic field lines from the location of 6 Re (Figure 5.9 (right [c]). This is the reason why the initial brightening of an arc at the time of substorm occurs at about 65° in latitude.

This place may be located just at the boundary or just outside the auroral oval, or a narrow belt between the southern boundary of the oval and the diffuse aurora, where the ionospheric conductivity is low.

Stage II: Flow of the disrupted cross-tail current

When an arc brightened, it immediately advances poleward with a speed of 250 m/s as shown in Chapter 4 or Figure 6.14. The auroral electrojet (*e-f* in Figure 6.10) advances closely with the poleward advancing of arcs in this expanded ionization belt (Akasofu and Chapman, 1962). This can explain a rapid growth of the UL current, because of the increased ionization belt by the poleward advancing arcs (Figures 6.4b and 6.14).

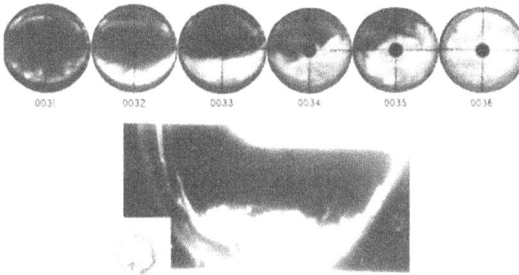

Figure 6.14 Upper: All-sky images of sudden brightening at the expansion phase onset (Farewell, Alaska, 63°; the arc was located at about 60°) and the subsequent rapid poleward advance of arcs (GI). Lower: DMSP image of the expansion phase (DMSP).

Thus, the highly ionized ionospheric belt thus produced allows the disrupted cross-tail current to flow into the ionosphere. The expanded ionization belt produces the circuit (*cefd*) in Figure 6.15 (right), which is Bostrom's azimuthal component (Figure 6.15: left).

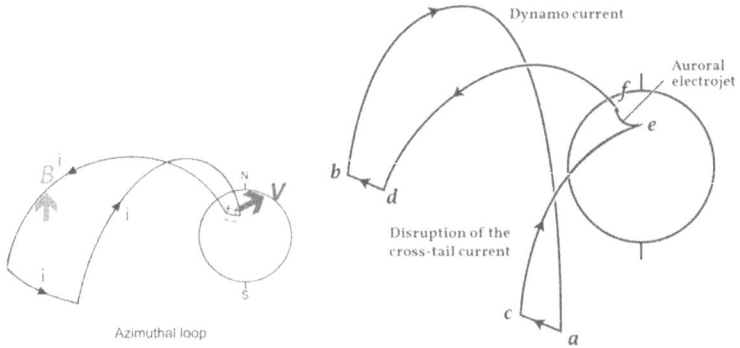

Figure 6.15 Left: The magnetic field **B** is produced by the current system, and **V** shows the poleward advance of the aurora. and earthward end of the circuit. Right: This is the same as Figure 6.10. It is a revised version of the left; the equatorial current of Bostrom's system is replaced by the disruption of the cross-tail current (c-d). The part (a-b) is the dynamo-induced current.

This azimuthal current produces a northward magnetic field **B**: Figure 6.15 (left). For a typical observed value of $B = 50$ nT (observed on the ground and at the ATS location), the total magnetic flux of the current system of Bostrom's current in Figure 6.15 (left) is about the same as the flux of the earth's dipole field of the area covered by the poleward advanced aurora. Thus, it is likely that the earthward end of the current system moves poleward as a result of its own field, about 500 km or about 5° in latitude as is commonly observed. Thus, the poleward advance of the aurora during the expansion phase can semi-quantitatively be explained by the poleward shift of its earthward end by its own field.

Another possibility is a possible tailward movement of the diversion of the cross-tail current (Akasofu, 1972). In fact, there is an interesting observation by Jacquey et al. (1993), which showed the tailward propagation of the disruption of cross-tail current.

Any theory of the expansion phase must explain the extent of the poleward expansion 300-500 km) of the auroral oval of several hundred kilometers (Figure 6.14). For very intense substorms, it can be 1000 km. Therefore, the poleward advance is not what the term 'dipolarization' implies; this term implies that the distorted dipole field before the expansion phase goes back to the dipole configuration; the aurora moves 500 km poleward from the original location before substorm onset (cf. McPherron et al., 1982).

(b) Ionospheric currents: Auroral electrojet

When the earthward electric field caused by the depolarization (Figure 6.11, red arrow) is transmitted to the ionosphere, it becomes a southward (equatorward)-oriented electric field, which drives the Hall current. The Hall current is the main current in the auroral electrojet; Figures 6.16 and 6.17.

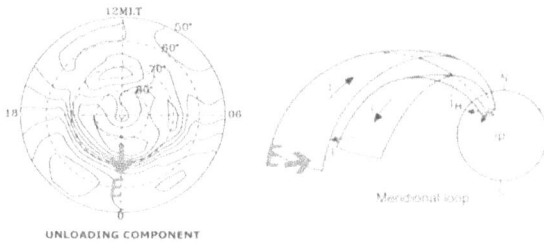

Figure 6.16 Left: the southward electric field (the transmitted earthward electric field on the equatorial plane) generates the UL current. Right: The meridional current proposed by Bostrom (1964). The Hall (J_H) and Pedersen (Jp) currents are shown.

The disrupted current (ef) flows as the Pedersen current, which generates the north-south polarization electric field in the expanded ionized belt, which in turn generates the Hall current. Thus, the westward current, *the auroral electrojet*, of 10^6 -10^7 A, is the combination of these currents (recorded as the UL current).

Thus, the resulting ionospheric current is complicated by the fact that the ionosphere has an anisotropic conductivity and also by the fact that the initially brightened arc advances poleward during the expansion phase, resulting in the expanded ionized (conductive) belt.

In order to see characteristics of both the Hall and Pedersen currents, Yasuhara et al. (1975) separated them from the total current based on the data from Alaska meridian chain of stations. They are shown in Figure 6.17.

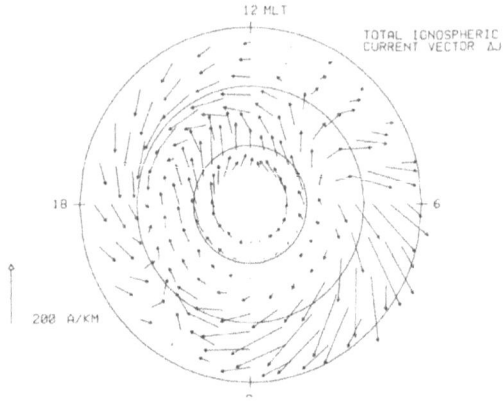

Figure 6.17a The distribution of the total current.

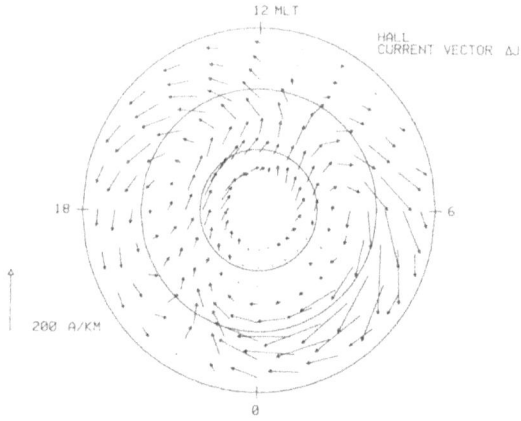

Figure 6.17b The distribution of the Hall current. It is separated out by the Hall conductivity.

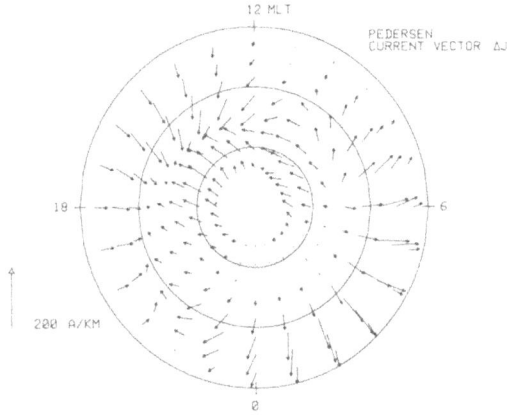

Figure 6.17c The distribution of the Pedersen current. It is
separated out by the Pedersen conductivity.

6.4 Double layer

Bostrom's current system shown in Figures 6.13 and 15 cannot work
without U-shaped electric potential structure called the "double layer" along
field-aligned currents; the whole circuit is open without the double layer. The
double layer can accelerate electrons downward to produce the ionization in
the ionosphere and the aurora (and can accelerate O^+ ions upward for the ring
current); the MHD approach does not allow such an electric field along
magnetic field lines; Figure 6.18.

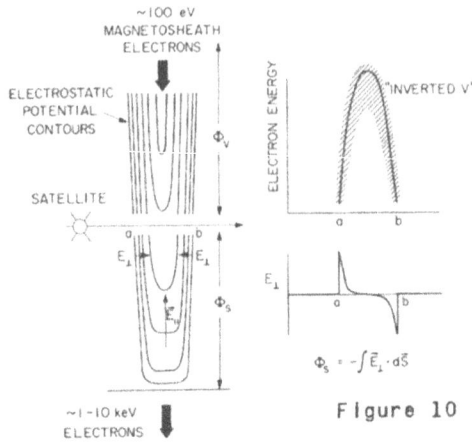

~100 eV
MAGNETOSHEATH
ELECTRONS

ELECTROSTATIC
POTENTIAL
CONTOURS

SATELLITE

"INVERTED V"

~1-10 keV
ELECTRONS

Figure 10

Figure 6.18 Left: Observed double layer, its potential structure (Don A. Gurnett). Right: The inverted V structure of precipitating electrons, supporting the double layer structure. See also Figure 1.15 for the ray structure of auroral arc.

The double layer is definitely an *observed fact,* so that the success of any auroral theories depends on how to produce an appropriate field-aligned current with the double layer.

There are a large number of satellite observations of the double layer associated with auroral field-aligned currents in the magnetosphere. The first observations have been conducted by satellites (cf. Gurnett, 1972; Mozer, 1973; Newell et al.,1996; Mozer and Hull, 2001).

In the earth's auroral conditions, various observed values related to the double layer are summarized by Karlsson et al. (2012, 2020): Field-aligned potential drops of the order of 6 KV or more, field-aligned currents of 10^{-1}-10^{1} $\mu A/m^{2}$, acceleration of magnetospheric electrons from 300 eV to 10 KeV and more, and the estimated thickness of the double layer 10 KV per 1 km, located between 0.5-2.0 Re above the ionosphere. Thus, it is certain that when field-aligned current is present, the double layer can be formed under a certain condition (see below) when its intensity of the current is high enough.

The magnetosphere is the only place so far to observe *in situ* field-aligned current with the double layer in space physics and astrophysics, so that its

168

importance has not necessarily been well recognized yet.

In general, *without an electric field along magnetic field lines*, it is extremely difficult to accelerate the current-carrying electrons along magnetic field lines to as high as 10 KeV from their original energy of 300 eV in the magnetosphere in order for them to penetrate down to the ionosphere; there have been many proposed electron acceleration processes in magnetized plasmas, but we are considering here the acceleration of current-carrying electrons along magnetic field lines by the formation of the double layer.

This is exactly one of the examples of what Alfven (1968) pointed out, namely "[if we neglect electric current], we deprive ourselves of the possibility of understanding some of the most important phenomena in cosmic plasma physics."

Therefore, without the field-aligned current with the double layer, both Bostrom's current systems cannot reach the ionosphere.

Alfven (1986) considered that the development of the double layer is associated with the nature of *electrical circuit in plasma*, so that it is necessary to consider its formation in terms of the *whole circuit problem,* as well as local processes. The circuit in a plasma tries to close itself.

The altitude range (0.5-2.0 Re) suggests that the location of the double layer is the transition region between the magnetosphere and the ionosphere (not far out in the magnetosphere); one of the reasons for the double layer to form there may be the need for the continuity of field-aligned electric current along the circuit (the whole circuit problem), when field-aligned currents encounter some discontinuity, such as non-uniformity of the conductivity (electron density) along magnetic field lines of the circuit; in this case, it is likely that the electron density transition from the magnetosphere to the ionosphere. The amount of the potential drop might depend on the conditions which require the current continuity.

This is the way how the two Bostrom's current system can ionize greatly the ionosphere (increasing the conductivity) *by the field-aligned currents with the double layer*. It is this process, which allows the primary M-I system to be able to dissipate the power as an ordinary current system during the recovery phase until the power subsides below about 10^{11} W; Figure 6 4.

The double layer can also accelerate O^+ ions upward; Figures 6.19 and 6.20. The upward accelerated O^+ ions by the double layer during the expansion phase is observed in the magnetosphere; then, they are sent back to toward the earth by the convective flow (Section 5.6) and form the ring current belt; Section 2.3 (c).

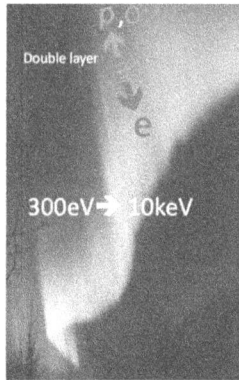

Figure 6.19 Double layer (=), accelerating electrons downward and protons/O^+ ions upward. It is likely that the double layer is located at a distance of 0.5 - 2 Re (Karlsson, 2012, 2020).

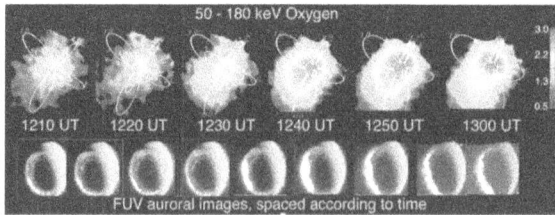

Figure 6.20 O^+ ions in the ring current belt at the time the expansion phase onset. In the all-sky camera record, the righthand side of each image is north; the poleward expansion is seen near the southern horizon (Fok).

6.5 Recovery phase

Figure 6.4 shows that after the expansion phase, the primary M-I system follows more or less the power until the power becomes less than 10^{11} W.

$$\text{Power (t)} \approx \text{Dissipation (t)}.$$

The significance of why the expansion phase and UL current occur at the *early phase* (40-100 minutes after the power is increase above 10^{11} W) of auroral substorms (lasting 2-3 hours) is that it is in this way that the Bostrom's two current systems remove a sort of 'anomaly' (the ionosphere is not conductive to receive the DD current) of the primary M-I system by the double layer, so that the primary M-I system can function properly after the expansion phase by the help of the secondary M-I system (the combination of the two Bostrom currents), namely the power is nearly equal to the dissipation rate as a function of time.

6.6 Summary of the synthesis work

A synthesis of various processes related to auroral substorms suggests the following sequence of the processes, including the explosive process:

1. *Power increase above 10^{11} W,*
2. *Blocking of the current flow in the ionosphere during the growth phase (because of its low conductivity),*
3. *Accumulation of power (energy) in the inner magnetosphere,*
4. *Occurrence of an instability of the cross-tail current, when the energy reaches about 10^{16} J,*
5. *Disruption of the cross-tail current/depolarization,*
6. *Growth of the earthward electric field by the dipolarization,*
7. *Establishment of the meridional sheet current system with field-aligned current and the double layer,*
8. *Ionization of the ionosphere by current-carrying electrons (sudden brightening of the arc), the expansion phase onset,*

9. *Magnetic field of the azimuthal component shifts its earthward end, causing the northward advance of arcs.*

10. *Widening of the ionization belt by the northward advance of arc (s),*

11. *Flow of the disrupted cross-tail current into the ionosphere, resulting in the auroral electrojet,*

12. *Dissipation of the accumulated energy as the Joule heat in the Ionosphere—the expansion phase,*

13. *After the expansion phase, the DD current follows, more or less, with the power. The primary M-I system functions normally.*

The above synthesis may be graphically shown in Figure 6.21.

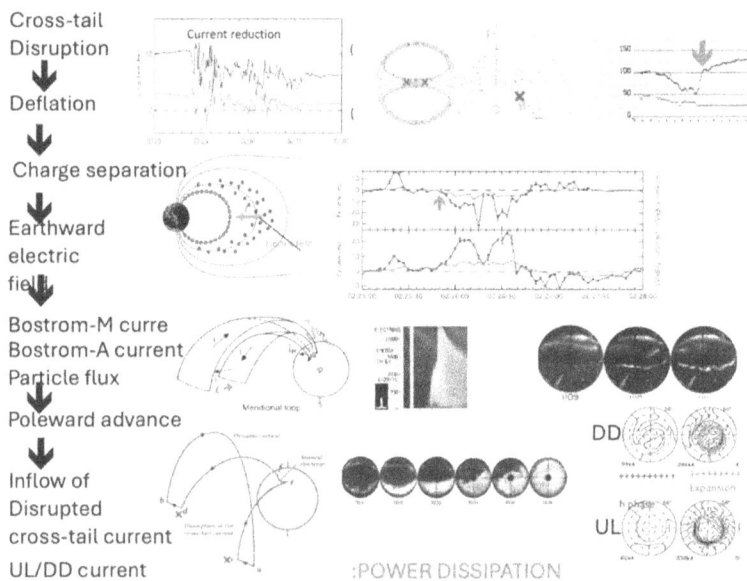

Figure 6.21 Synthesis of the expansion phase in a graphic form.

This view of the expansion phase may be supported by:

(1) The expansion phase occurs only at an early phase of substorms;40-100 minutes after the power begins to increase.

(2) A large expansion occurs *only once* (even a constant high power lasts for 3-4 hours), unless there occurs a new power increase.

(3) The amount of energy spent during the expansion phase is about the same as the accumulated energy during the growth phase.

(4) The duration of both the growth phase and the expansion phase is about the same (40-100 minutes).

(5) The recovery phase remains strong even after the expansion phase so long as the power remains above 10^{11} W.

In the course of our study, we learned that each auroral phase is caused by a specific sequence of physical processes. This understanding has led us to describe each phase during a substorm. As a result, the past morphological terms may be changed. For example:

The growth phase as the energy accumulation phase,

The expansion phase as the explosive energy release phase,

The recovery phase as the normal functioning phase.

Actually, Yasha Feldstein suggested to me once that the growth phase should be called the "preparation phase [for the expansion phase]." In fact, the name of the three phases has been based on morphological features.

Therefore, once the physics of these phenomena becomes eventually understood, new names based on physics should be considered for the three phases (or new phases).

Individual substorms are far more complicated than what are described in the above for the average (medium intensity) auroral substorm. Individual substorms are controlled by (1) how IMF varies as a function of time (not like a step-function at onset) and how it varies in time, (2) ionospheric conditions (the degree of ionization before the onset and after the expansion phase, as well as during the expansion phase), and (3) how the power varies after the expansion phase and many others as many factors. It is known that in some cases the expansion phase occurs at the time of storm sudden commencement [ssc] (Tsurutani, and Hajra, 2023); in such cases, the IMF Bz component is often negative at about the time of geomagnetic storm onset (ssc).

In this chapter, we did not make a review on the work based on the magnetic field line approach. Substorm research seems to be progressing well (Knudsen et al., 2021).

6.7 Identification of individual auroral displays in terms of physical terms

In this section, I attempt to describe various types of auroral displays in terms of physical processes; see also (Akasofu, 2012).

(a) Initial brightening

The initial brightening was described in detail in Section 6.2, but it is briefly summarized here. When the dipolarization occurs as the result of the reduction of the cross-tail current, the inner magnetosphere is deflated. The deflated field lines carry electrons earthward, but not protons.

Thus, a charge separation is expected to occur (resulting in the *earthward* electric field *E*). The shifted electrons are discharged along magnetic field line to the ionosphere. It is in this process the double layer accelerates these electrons.

As a result, the sudden brightening of an arc at the expansion phase onset occurs at the equatorward boundary of the auroral oval, which corresponds to the intersection line between the outer boundary of the outer radiation belt and the ionosphere (Sections 3.3 and 4.3). This simple fact alone indicates that the main cause of auroral substorms is located in the inner magnetosphere, not in the magnetotail.

(b) Poleward advance

The secondary M-I system produces an upward magnetic field *B*, which can shift the earthward end of Bostrom's current systems. This causes the most spectacular feature among various auroral activities, the explosive poleward advance of auroral arcs and the auroral electrojet soon after the initial brightening. The front of the exploding arcs can reach as far as 78° from 65° (Craven et al.,1984). The speed of the advance is about 250 m/s; Figure 6.22.

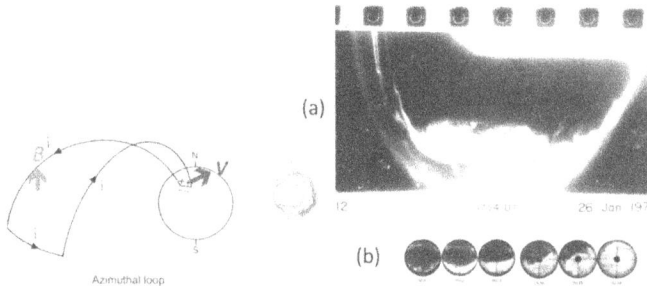

Figure 6.22 Left Azimuthal component of Bostrom's current system (Bostrom, 1964). It produces an upward magnetic field **B**. Thus, the earthward end of the UL current moves toward higher latitude with its own magnetic field. **V** indicate the poleward movement of the earthward end of the current system.: Right: (a) DMSP image. (b) All-sky images, showing the poleward expansion of the auroral oval (GI).

Figure 6.23 shows all-sky images along the Alaska meridian chain of all-sky images. The initial brightening arc was located a little poleward of College (Fairbanks), which was also seen in the equatorward sky of Fort Yukon. There was no specific auroral activity to the poleward of College at the time of the initial brightening, except there was some arcs poleward of Mould Bay.

Another important feature in Figure 6.24a is that the brightened arc advanced poleward immediately after the initial brightening; in this case, the poleward advance reached as far as Sachs Harbor, expansion range being about 700 km.

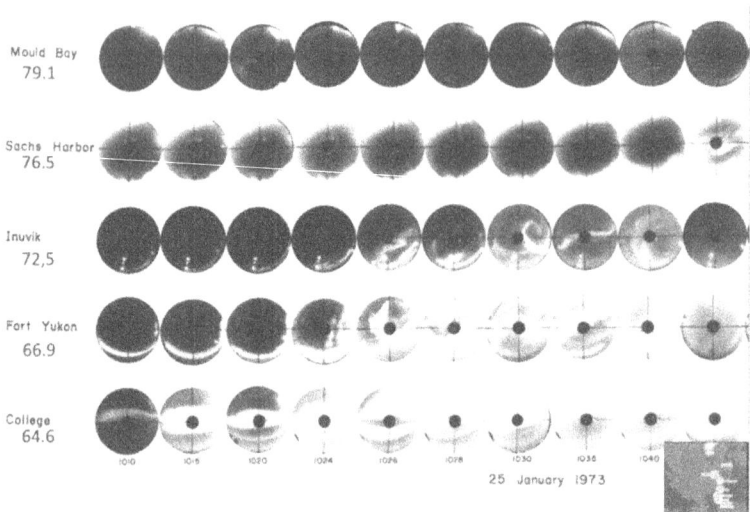

Figure 6.23a All-sky camera records from the Alaska meridian chain of stations. The initial brightening arc was located a little poleward of College (Fairbanks). and began to advance poleward at 10:24 UT and reached Sachs Harbor at 11:50 UT. The Alaska meridian chain of all-sky cameras is also shown (University of Alaska Fairbanks).

Figure 6.23b shows an example of great expansion phases, which occurred during the great geomagnetic storm of February 11, 1958. It shows both all-sky images at College (Fairbanks) and geographical extent recorded by the IGY all-sky camera network; the speed of poleward advance was more than 1 km/s. The auroral activity expanded over the latitudinal extent of about 20° in a matter of 30 minutes over the whole US sector (Akasofu and Chapman, 1963).

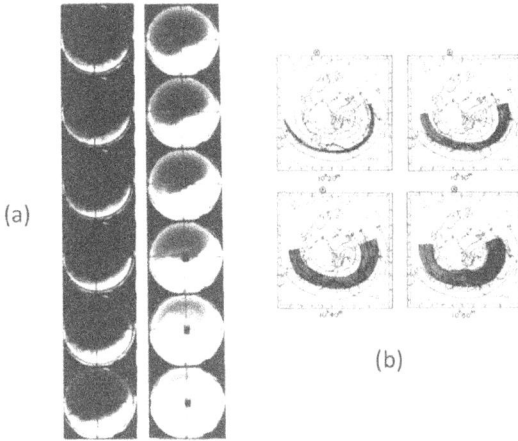

Figure 6.23b An example of a large poleward expansion phase recorded by the IGY auroral network of all-sky cameras on February 11, 1958. (a) College (Fairbanks) all-sky images, one minute apart (GI). (b) The expansion occurred over the whole US sector (Chapman and Akasofu, 1963).

(c) Westward traveling surges (WTSs)

The most prominent display in the evening sky is westward traveling surges (Figure 6.24). They propagate toward the day sector along the auroral oval and are accompanied by the westward extension of the auroral electrojet (*f* point of Figure 6.15); It is likely that their westward advance is caused by the westward development of diversion of the cross-tail current (the *d* point of Figure 6.15).

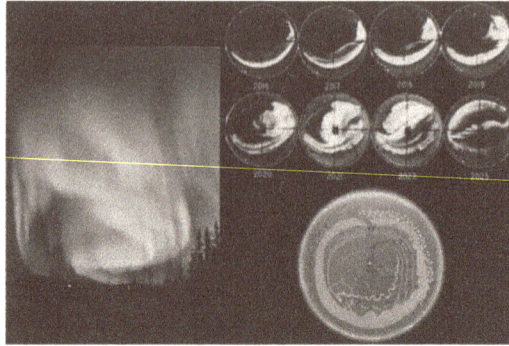

Figure 6.24 Westward traveling surge; photograph and all-sky images (not simultaneous), (GI) and an illustration of westward traveling surge (circled).

Westward traveling surges show often a counterclockwise motion, suggesting a positive space charge (Figure 6.25). It is likely related to the fact that it is the westward end of the auroral electrojet.

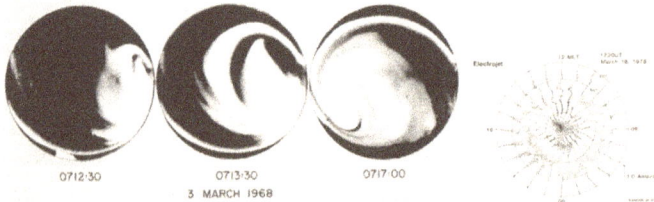

Figure 6.25 Left: An example of westward traveling surges. Note a counterclockwise motion arc (NASA; Airborne auroral expedition). Right: Auroral electrojet. The surge is near the westward front of the auroral electrojet.

When a WTS advances along the auroral oval in the evening sector, a distinct positive H component change of the magnetic field is recorded a little *south of the oval*. The occurrence of WTSs in the evening sky indicates that a substorm is already in progress in the midnight sector.

This positive H component observation (often designated as DP2) is often considered as the growth phase feature, but it is not.

During a substorm, the DD component is a two-cell current (Figure 6.26), and in the afternoon-evening sector, it has the eastward current in the lower side of the auroral oval, which causes a positive H component. Thus, the H component increase is a part of substorm field, not the precursor of substorm.

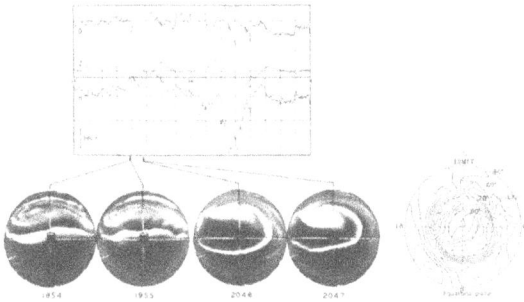

Figure 6.26 An example of westward traveling surge, recorded at Fort Yukon, about 200 km north of Fairbanks (February 18, 1958); (GI). It shows also the College (Fairbanks) magnetic record, showing a prominent positive change in the H (bottom) component. The insert is the DD current. The positive H component is produced by the eastward electric current of the eddy DD current in the afternoon or evening sector.

Thus, when a time series of the AE index is divided in terms of a large decrease as T = 0 and assembled, such a positive change appears *before* T = 0. However, when day-to-day magnetic records with all-sky camera images are examined, it is clear that the positive change occurs in the evening sector, when a substorm is already in progress in the midnight sector.

(d) Patches

One of the displays in the morning sector is disintegration of the arc structure; Figure 6.27. This occurs in the southern part of the auroral oval. It appears like cumulus clouds. However, they are not. They are disintegrated arcs with the vertical structure (with the ray structure). This was confirmed by flying across (N/S) the auroral oval a few times by the NASA Galileo flight.

Patches drift eastward with a speed of about 300 m/s. The plasma flow (E x B drift) associated with the auroral electrojet seems to be very turbulent in the morning sector (no observed report by satellites).

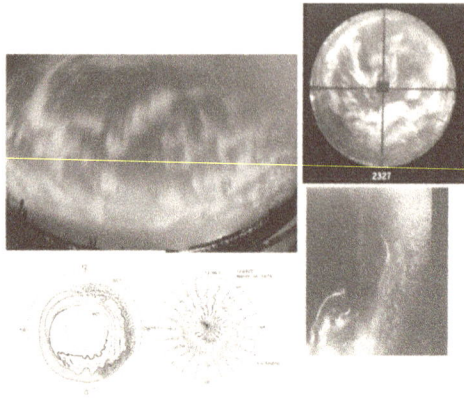

Figure 6.27 Patches, visual image, all-sky image (GI) and DMSP image (DMSP). The patches are actually not patches , but are disintegrated curtain-lie arcs.

(e) Omega bands

In the northern part of the oval, an auroral arc tends to show inverted omega-type distortion. It drifts eastward; Figure 6.28.

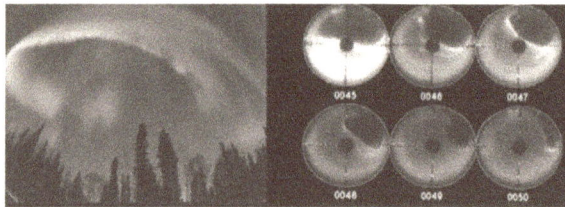

Figure 6.28 Example of the Omega bands, visual and all-sky images (GI)

(f) Torch

In the equator side of the drifting patches, the diffuse aurora develops a very large-scale wavy structure (torches); Figure 6.29. Since the diffuse aurora seems to be caused by the outer radiation belt, their location and size suggest a very large-scale (a few earth radii on the equatorial plane) deformation of the radiation belt. This was studied by Forsyth et al, (2021).

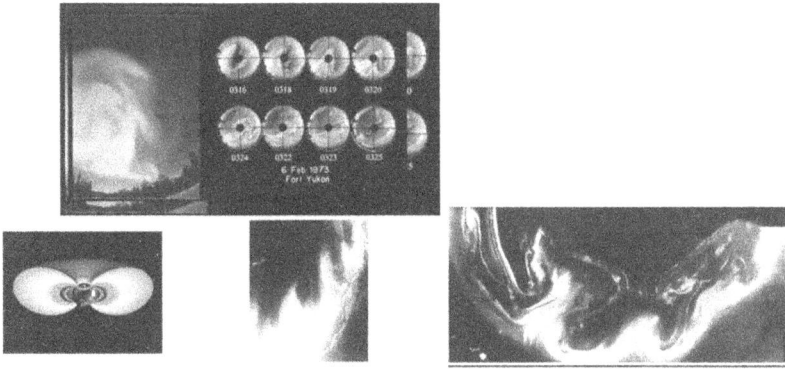

Figure 6.29 Upper: Example of torches, visual and all-sky images (GI). Lower: Van Allen outer radiation belt and two DMSP images.

Recently, I found a book "Auroral Physics" in the GI Library (Knudsen et al. 2021). I was so pleased that I wrote to David Knudsen, the chief editor. I am very pleased to find that auroral research is nicely assembled in the book "Auroral Physics". "I am pleased to see your book and wish further advances by all the contributors. In my days, the general trend was such that all auroral processes had to be based on magnetic reconnection".

He immediately replies:

"Dear Syun So nice to hear you and thank you for your positive comments. I'm sure to recognize the artwork on the front cover as inspired by your well-known published figure. Our main goal in proposing that book was to highlight unanswered questions. ----".

In my days, the general trend was such that all auroral processes had to be based on magnetic reconnection. I mentioned also my point in a paper in the AGU book edited by Keiling. I had a talk with Eric Donovan, co-editor, saying that the research direction has to be changed, and he agreed with me.

Two recent edited books on substorms are Keiling et al. (2012). and Knudsen at al. (2021).

6.8 Relationship between the main phase of geomagnetic storms (the ring current) and auroral substorms

It is worthwhile to look back how far we have reached in studying the main phase of geomagnetic storms and the formation of the ring current. This was the first problem Chapman asked to me to consider, when I met him first; Figure 6.30.

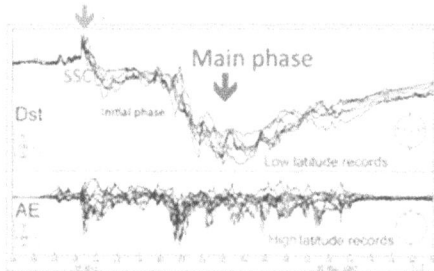

Figure 6.30 A typical example of geomagnetic storms

At that time, I had no idea that the main phase was related to the aurora. Now, we know at least that auroral substorms play a crucial role of the cause of the main phase. The field-aligned current generates the double layer. The double layer can accelerate current-carrying electrons downward to produce the aurora and also accelerate O^+ ions upward. The O^+ ions are eventually brought to the ring current belt; Figure 6.31.

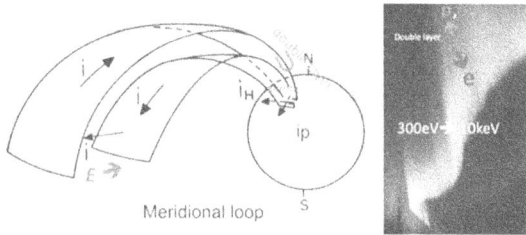

Figure 6.31 Bostrom's meridional current system and the double layer. The double layer accelerates current-carrying electrons downward and accelerating O^+ ions upward.

It is quite likely that the accelerated O^+ ions and protons are injected to the magnetotail and then flow into the ring current belt in the midnight sector during the expansion phase of auroral substorms; Figure 6.32.

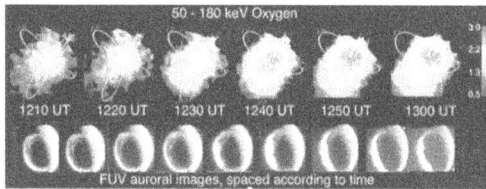

Figure 6.32 Simultaneous observation of O^+ ions in the ring current and auroral substorm (all-sky camera images); (Fok, 2003).

Indeed, the UL current has a well-defined relationship with the main phase decrease (Dst decrease) more than the DD component, although the DD component contribute also, perhaps through the convection toward the ring current; Figure 6.33.

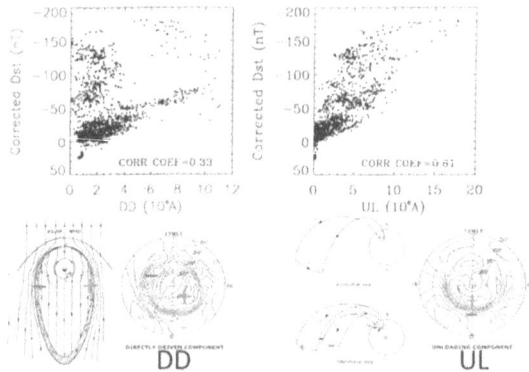

Figure 6.33 Upper left: Relationship between DD current intensity and the Dst index. Upper right: Relationship between the UL current intensity and the Dst index. Lower left. The magnetotail convection and the resulting DD current. The Bostrom, current and the UL current.

All together, Chapman's problem to me, the cause of the main phase, may be solved, if we understand more about auroral processes.

6.9 Remarks at the end of my substorm research

In this chapter, we could succeed tentatively in understanding why auroral substorms have the explosive feature as the sequence of 13 processes. This is just one possible sequence of observed facts.

After all, it has taken more than 60 years to reach this far. The aurora seems to be infinitely mysterious.

This is as far as I could reach on the cause of auroral substorms and the formation of the ring current (the main phase of geomagnetic storms) by the combined efforts of a study of auroral substorms and geomagnetic storms. I owe a large number of researchers in magnetospheric physics and space physics in order to come to this far. It is hoped the younger generations will advance further substorm study. This is the main reason for writing this book.

If my 'story' proceeded as I described somewhat in a textbook way (although I tried to avoid), my research life would have been much simpler. On many occasions, I lost the direction of the approach in the sequence (what I should consider in advancing ahead). Some of the steps were earlier and some others were later in each part of the sequence. Each step of the sequence took a few days, many months or many years, depending on each step. Thus, it has taken 60 years in reaching the present understanding.

What is described in this chapter is *an example* of *morphological theory of auroral substorms* of medium intensity. Hopefully, a solid *mathematical* theory of auroral substorms will emerge on this basis or a new morphological study in the future. Particularly, I encourage young researchers to establish a new theory of auroral substorms based on my morphological theory or by others.

In order to stress this point, in Chapter 8 I emphasized that any worthwhile mathematical theory must be based on a solid morphological theory. It is my hope that the morphological theory described in this chapter will be useful.

My motive

The particular reason for the study in this chapter is that the substorm research has come to the point, in which it is necessary to begin to assemble observed facts and examine how these facts can be synthesized to find how auroral substorms occur in the way they occur.

Such a synthesis work would be useful for further development of auroral substorms by knowing what processes may be missing.

In this chapter, we showed an example of such a study. Many researchers are still too busy in studying their individual processes and have no time to consider how their processes are related to each other.

The first meeting with Hannes Alfven was most crucial for me in this chapter.

He suggested strongly that I should take the electric current approach, in spite of overwhelmingly popular magnetic field line approach.

In this chapter, we could succeed partially in understanding why auroral substorms have the explosive feature as the sequence of 13 processes, instead of relying on the presumed magnetic reconnection.

It is hoped that researchers in space physics recognize at least that there is another approach, other than the magnetic field line approach. It is for this reason I contrasted both approaches wherever possible.

Episodes

(1) Aurora on Jupiter and Saturn

During our auroral research was in progress, we had been waiting for the news about the aurora on Jupiter, since it is a magnetized planet and thus we expected auroral reports. We were very happy to see images of the aurora on Jupiter first. Further, it was a great surprise that in addition to the presence of the aurora, the images seem to show an auroral oval-like distribution. Furthermore, they showed also substorm-like features (the poleward expansion). These findings were more clearly demonstrated later on Saturn.

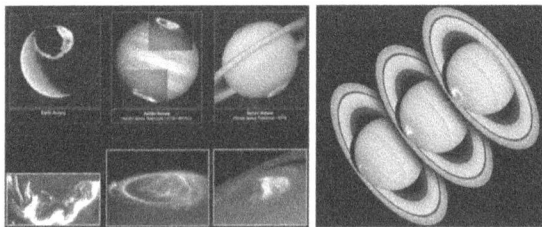

Left: the aurora on the three magnetized plants (NASA). Right: A substorm-like feature on Saturn (Boston University).

(2) Aurora on exo-solar planets

When astrophysicists found stars with planets and tried to find ways to detect earth-like life there, I thought that one way may be to look for oxygen emissions in their aurora (Akasofu, 1999). If their atmosphere has oxygen, there may be at least plants.

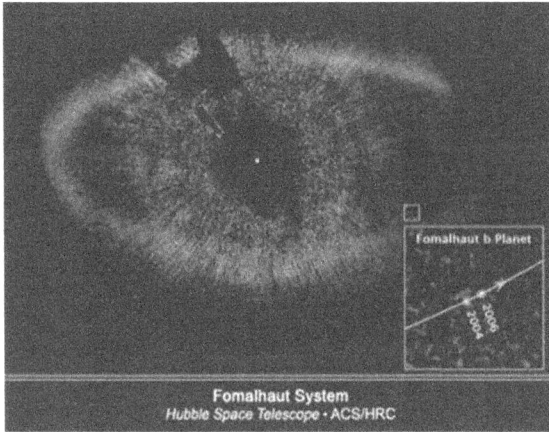

Fomalhaut System
Hubble Space Telescope · ACS/HRC

`Q`QAThe presence of exo-solar planets (Kalas et al., 2008).

(3) Alfven's visit to Chapman's bust

In September 1974, I invited Hannes Alfven to the Geophysical Institute of the University of Alaska Fairbanks. When he learned of Sydney Chapman's bust at the Chapman Building, he wanted to take a photograph together with me there. From 1951, they had a long debate, more than 20 years, on the theory of geomagnetic storms and the aurora (Akasofu, 2003). I thought that his wish to see Chapman's bust is an indication that he was happy to have had the long debate with Chapman. I was very fortunate to be able to learn from both (Chapter 2 from Chapman; Chapters 5 and 6 from Alfven). I feel that I tried to combine their theories by considering that the aurora occurs as an electrical discharge (Alfven) within the magnetosphere (Chapman).

With Hannes Alfven at the front of Chapman's bust in the Chapman Building (GI).

(4) Second book on substorms

Meanwhile, I published the second book from the same publisher in 1977, titled *Physics of Magnetospheric Substorms*.; Akasofu (1977). Compared with the first book in 1968, we can see how much this field had developed in less than 10 years.

References

Akasofu, S.-I., 1972, Magnetospheric substorms: A model-1970, *Solar-Terrest. Phys.: Proceedings of the International Symposium on Solar-Terrestrial Physics held in Leningrad, U.S.S.R. 12-19 May 1970, Part III*, Dyer, E.R. and J. R. Roederer (eds.), 131-151, D. Reidel Pub. Co., Dordrecht,.

Akasofu, S.-I., 1999,Auroral spectra as a tool for detecting extraterrestrial life, *Eos, 80*, 397.

Akasofu, S.-I., 2012, Auroral Morphology; A historical account and major auroral features during auroral substorms, 29-38, in *Auroral Phenomenology and Magnetospheric Processes, Earth and other planets*, ed by A. Keiling, E. Donovan, F. Bangenal and T. Karlsson, Geophysical Monograph **197**, American Geophysical Union, Washington D.C.

Akasofu, S.-I., 2017a, Where is the magnetic energy for the expansion phase of auroral substorms accumulated? 2. The main body, not the magnetotail, J. Geophys. Res.,**122**, 8479, doi:10.1002/2016JA023074

Akasofu, S.-I., 2017b, Auroral substorms: Search for processes causing the expansion phase in terms of the electric current approach, Space Sci. Rev., **212**,341. Doi: 10. 1007/s11214-017-0363-7

Akasofu, S.-I., Auroral spectra as a tool for detecting extraterrestrial life, *Eos, 80*, 397, 1999.

Akasofu, S.-I., Lui, A. T.Y. and Lee, C.-H. Synthesizing auroral substorms processes based on magnetosphere-ionosphere electric currents, Front.Astron. Space Sci., 12:1521520, doi: 10.3389/fspas.2025.1521520

Akasofu, S.-I. 2023, A new understanding of why the aurora has explosive characteristics, Mon. Not. Roy. Astronom. Soc., **518**, 3286, https://doi.org/10.1093/mnras/stac3187

Akasofu, S.-I. and Chapman, S., 1962, Large-scale auroral motions and polar magnetic disturbances--III: The aurora and magnetic storm of 11 February 1958, *J. Atmos. Terr. Phys., 24*, 785.

Alfven. H., 1968, The second approach to cosmical electrodynamics,in *The Birkeland Symposium on Autora and Magnetic storm,* ed. byA. Egeland and J. Holt, Centre National de la Recherche Scientifique, Paris, 439.

Alfven, H.,1981, *Cosmic Plasma*, D. Reidel Pub. Co.Dordrecht-Holland.

Alfven, H.,1986, Double layers and circuits, IEEE, **PA-14**, No.6,779.

Alfven, H., *1967*, Birkeland Symposium on Autora and Magnetic storm, ed. byA. Egeland and J. Holt, Centre National de la Recherche Scientifique, Paris, 439.

Bostrom, R., 1964, A model of the auroral electrojets, J. Geophys. Res. **69**, 4983.

Deforest, S. E. and McIlwain, C. E., 1971, Plasma clouds in the magnetosphere, J. Geophys. Res., **76**, 3587.

Fok, M. C. et al.,2006, Impulsive enhancement of oxygen ions during substorms, J. Geophys. Res.,**111**, A10222 doi:1029/2006JA011839

Forsyth, C., Sergeev, V. A., Henderson, M. G., Noshimura, Y., and Gallardo-Lacourt, 2020, Physical processes of meso-scale, dynamic auroral forms, Space Sci. Rev. **216:46**, https://doi.org/10.1007/s11214-020-00665-y

Gurnett, D. A., 1972, Electric and plasma observation in the magnetosphere,123-138, in Critical Problems of Magneospheric Physics, Proceedings of the joint COSPAR/IAGA/URSI Symposium, Spain, 11-13 May 1972.

Jacquey, C., Sauvaud, J.A., Dandours, J. and Korth, A., 1993, Tailward propagating cross-tail current disruption and dynamics of ear earth tail:a multi-point measurement and analysis, Geophys. Res. Lett., **20**, 983.

Keiling, A., Donovan,E., Bangenal, F and Karlsson, K., 2012, *Auroral Phenomenology and Magnetospheric Processes*: Earth and Other Planets, AGU, **197**, Washington, DC.
https://doi.org/10.1029/2011GM001179

Karlsson, T., 2012, The acceleration region of stable auroral arcs,227-237, in Auroral Phenomenology and Magnetospheric Processes: Earth and Other Planets, Geophysical Monograph Series **197**, AGU, Washington, DC.

Karlsson, T. et al., 2020, Quiet, Discrete auroral arcs-Observations, Space Sci. Rev. **216: 16**, https://doi.10.1007/s11214-020-0641-7

Knudsen, D. J., Borovsky, J. E., Karlsson, T. and Kataoka, R., 2021, *Auroral Physics*, Springer.

Lui, A. T. Y., Chang, C.-L., Mankofsky, A., Wong, H.-H. and Winske, D., 1991, A cross-field current instability for substorm expansions, J. Geophys. Res., **96**, 11389, https://doi.org/10.1029/91JA00892

Lui, A. T. Y. and Kamide, Y., 2003, A fresh perspective of the substorm current system and its dynamics, Geophys. Res. Lett., **30**, (18),1958, doi: 10. 1019/2003GL017835

Lui, A. T. Y., 2011, Reduction of the cross-tail current during near-erath depolarization with multisatellite observations, J. Geophys. Res., **116**, A12239: Doi: 10. 1029/2011AJA017107

Lui, T. T. Y., 2020, Evaluation oft he cross-field current instability as a substorm onset process with auroral bead properties, J. Geopys. Res., doi: 1029/2020JA027867.

McPherron, R. L., Russell, C. T., and Anlory, M. P.,1983, Satellite studies of magnetospheric substrom on August 15,1968: A phenomenological model for substorms, J. Geophys. Res.,**78**, 3131.

Mozer, F. S., 1973, Analyses of techniques for measuring DC and AC electric fields in the magnetosphere, Space Sci. Rev.,14, 272. https://doi.org/10.1007/BF02432099

Mozer, F. S. and Hull, A., 2001, Origin and geometry of upward electric fields in the auroral acceleration region, J. Geophys. Res. **106**, 576. **DOI:** 10.1029/2000JA900117

Newell, P. T., Lyons, K. M. and Meng, C.-I., 1996, A large survey of electron acceleration events, J. Geophys. Res., **101**, 2599. https://doi.org/10.1029/95JA03147

Olson, W. P., 1984, Introduction to the topology of the magnetospheric current systems, 49, in *Magnetospheric Currents*, ed. By T. A. Potemra, AGU Monograph vol. **28.**

Sun, W., Xu, S.-Y., and Akasofu, S.-I., 2000, Mathematical separation of directly driven and unloading components in the ionospheric equivalent current during substorms, J. Geophys. Res.,**103**, 11695.

Tsurutani, B and Hajra, R., 2023, Energetics of shoch-triggered supersubstorms (SML,-2500 nT), ApJ., 946: 17.

Yasuhara, F., Kamide, Y. and Akasofu, S.-I., 1975, Field-aligned and ionospheric currents, *Planet. Space Sci., 23*, 1,355.

Chapter 7 Solar flares, Solar corona, Solar wind and Sunspots: Long-standing problems

The field of solar physics has several long-standing problems. They are the energy source of solar flares, the temperature of the corona, the cause of the solar wind and the formation of sunspots among others.

Here, we take the electric current approach in these long-standing problems. These problems are extremely difficult, but it is unusual that there is no generally acceptable theory (except sunspots) or reasonable understandings on most of these problems. One of the reasons seems to be almost complete lack of the concept of *electric current* in considering these problems. Thus, it is hoped that the electric current approach may provide a new insight into them. My study in this chapter is very qualitive, not quantitative. At the present time, we are only at the beginning of the electric current approach. Hopefully, a quantitative electric current approach will come later.

In Section 7.1, we consider a photospheric dynamo process for the power of solar flares, although magnetic reconnection has almost exclusively been considered as the power supply in solar physics. My criticism on magnetic reconnection is given in 7.1 (vi).

The high temperature of the corona was discovered in 1945, but all the proposed theories of heating the corona seem to reproduce the '*observed*' (based on Fe^{XII}) temperature (Aschwanden, 2005). I consider the possibility of ionization in coronal loops by current-carrying electrons in Section 7.2, because the ionization of the corona requires much higher energy electrons than the ionization of FeX^{II}; this is true in any theory of the coronal ionization (similar to the auroral ionization). Such a possibility has not been considered before. It is known that the loop structure of the corona is brighter compared with the surrounding. This is the starting point.

The sun blows out its uppermost atmosphere with a speed of 800 km/s all the way to the outer edge of the heliosphere (100 au). Gene Parker wrote the first paper on the cause of the solar wind in 1958 and coined the term "solar wind." However, its cause is still a matter of intense debate (cf. Viall and Borovsky, 2020). A new idea on the generation based on the powerful ($J \times B$) force is proposed in Subsection 7.3, but it cannot attain the speed of 800 km/s at the earth, only 200 km/s.

Sunspots are perhaps the solar phenomenon that has been studied for the longest time. However, we do not understand what sunspots are. The theory proposed by Babcock (1961) is still generally accepted, but has several obvious contradicting observations such as single spots (instead of a pair of spots). Thus, I propose a new idea, based on single spots in Section 7.4.

Although these are extremely difficult problems, it seems, nevertheless, that there is something fundamentally missing in solar physics, in spite of so many observational and theoretical efforts made on the above subjects in the past. As mentioned earlier, the fundamental problem in the above four problems in the past study is that *the concept of electric current is almost missing in their treatments, not more new observations.*

This reason for avoiding the concept of electric current seems to be partly due to the initial great success of magnetohydrodynamics (MHD), introduced by Alfven (1950) in his book *"Comical Electrodynamics",* in understanding various astrophysical phenomenon; in MHD, current *i* is replaced by *curl B*. He stated that many problems solvable by MHD ('froze-in' field line) had been solved, and the electric current approach may be needed for unsolved problems., but no one has listened to him. In auroral substorms, we learned that MHD brakes down at the critical points in the sequence of processes.

Another problem is that the intensity of electric current cannot be measured directly, but the situation is the same in auroral substorm studies. It can be overcome as magnetospheric physics has done (Chapters 5 and 6).

Further, it is shown in this chapter that some of old facts and data are actually more useful than many recent satellite data in understanding basics of solar flares and sunspots. For example, *spotless* flares (almost forgotten) seem to tell much more basics of solar flares than spectacular analyses in recent flare studies. *Single spots* (almost disregarded) seem to tell much more than recent detail observations about sunspots.

Thus, it may be worthwhile to go back to early fundamental problems, in addition to asking for higher space and time resolution studies. In this chapter, we encounter several examples in this regard. In this sense, going back could become going forward. For example, it may be that one of reasons why solar flares has remained as an unsolved problem for a long time is that the source energy for solar flares is considered to be already solved, but it seems that the process of magnetic reconnection is still uncertain (Section 7.1 [vi]). Unless the nature of magnetic reconnection and thus the basis of the source of power is uncertain or wrong, the whole processes of solar flares may need reconsideration. Thus, we go back to the abandoned photospheric dynamo.

7.1 Solar flares

This is a typical example to demonstrate that the electric current approach might provide a new way to study solar flares. First of all, we consider a photospheric dynamo as power supply, instead of magnetic reconnection based on the magnetic field line approach.

Solar flares are an "explosive" or a transient process. The total released energy is about 10^{25} J (10^{32} ergs). The cause of solar flares is one of the major problems in solar physics. The present flare studies depend on "explosive" annihilation of an anti-parallel magnetic field, magnetic reconnection, which is still "illusive" (Hesse and Cassak, 2020). How can one understand solar flares without knowing the power supply?

In my study of auroral substorms, I learned that the auroral "explosive" process requires first accumulation of energy and then its sudden release by a sequence of processes, but not by magnetic reconnection.

Thus, it may be possible to study solar flares in a similar way, although the power involved differs as much as 10^{10}.

I corresponded with Gene Parker in the analogy between auroral substorms and solar flares by saying that a photospheric dynamo (based on the same power equation of the auroral dynamo, but with a set of solar parameters) is found to be promising for solar flares. In one of the letters I received, he mentioned (the date missing):

"Dear Syun Your suggestion that substorms and flares are a consequence of an increased level of dissipation of magnetic energy above some critical level is an interesting thought. Your evidence for the substorm [by a dynamo process] is persuasive, at least tentative. It is more difficult with the solar flare because the rates can be $10**4$ L_s or more during a big flare---[more his explanations. Gene".

His concern was about the difference of the rate and the total energy, but we can show in the next section that our photospheric dynamo can generate enough power, which is based on the *observed* plasma speed and magnetic field. Further, the power can be accumulated for several hours before it is released, so that we can solve this problem of 10^{10}.

For these reasons, I thought I should try to study solar flares in the way I studied auroral substorms on the basis of electric current approach, in spite of a great difference of the magnitude of power and energy 10^{10}.

In fact, both solar flares and auroral substorm are morphologically similar as can be seen in the following, so that basic physics must be the same (electromagnetic energy dissipation phenomenon).

Morphologically, both solar flares and auroral substorms are well characterized by their atmospheric emissions of the curtain-like or ribbon-like form (the hydrogen emission [Hα]) of solar flares and the oxygen emission (55.77 nm) of auroral substorms. They are relatively short-lived. Both are associated with X-ray emission, production of high energy particles and radio wave emissions.

Solar flares	Auroral substorms
Hα emission	Oxygen emission
Type I & II radio waves	Kilometric radio waves (AKR)
X-rays	X-rays
Subcosmic-ray particles	150 KeV particles (O^+)

In 1984, Harold Zirin, a well-known and respected solar physicist. He and I ran movie films of both solar flares and auroral substorms together side by side in his office of the California Institute of Technology. At the end, Harold insisted that "auroral substorms are earth's *flares*," while I claimed: "solar flares are the sun's *auroras*, or solar substorms."

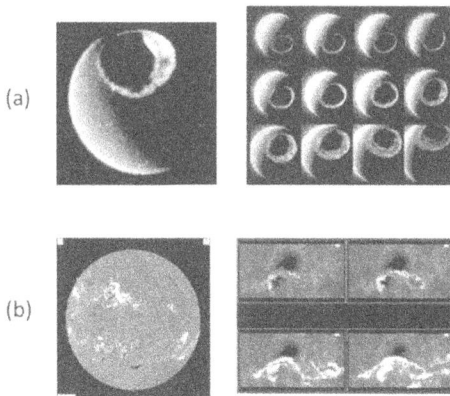

(a)

(b)

Figure 7.1 Auroral substorm (L. Frank) and solar flare (K. Shibata). It shows a large-scale view and their progress in time.

(a) Spotless flares

Since intense flares tend to occur in an active sunspot group (Figure 7.2), the whole situation is very complicated to describe and is difficult to comprehend, but those are the ones which are mostly studied and reported in term of magnetic reconnection.

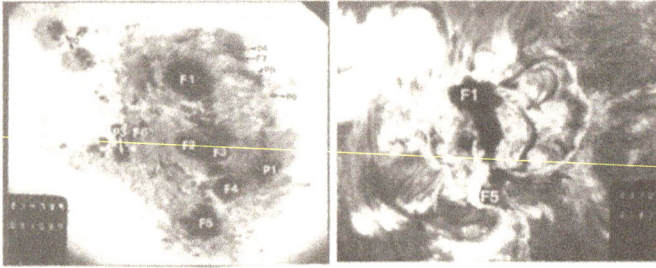

Figure 7.2 Solar flares in a very active sunspot group. Left: the distribution of sunspots. Right: Solar flares in the sunspot group (The Big Bear Observatory).

On the other hand, the simplest flares, called '*spotless flares*', have hardly been reported in recent years. However, I found that those simplest flares can tell us the basics of flare processes.

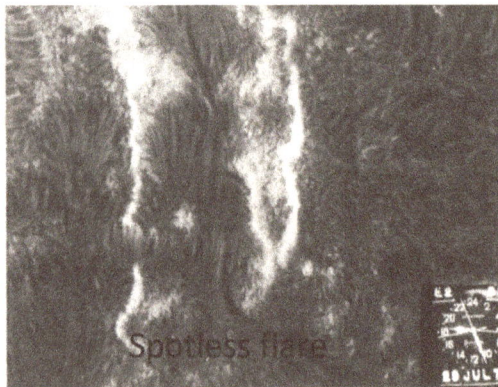

Figure 7.3 An example of spotless flares (Svestka, 1976).

Spotless flares occur *without* any sunspot around them (Dodson and Hedeman, 1970). They are the weakest, but just show the basic pattern of two-ribbon Hα emission often without any other interesting features. In fact, the two-ribbon emission is the largest total energy dissipation (Svestka, 1976); sub-cosmic ray particles are very energetic, but the total energy is much smaller than that of the two-ribbon emission.

First of all, *spotless flares indicate that sunspots are not needed as the direct cause of solar flares*; intense flares tend to occur within an active sunspot group (Figure 7.2), but the presence of spotless flares indicates that sunspots are not the direct cause of flares (such as a collision of two sunspots).

Thus, the **simplest** *type of flare is the one which has the very basic process of solar flare, particularly in terms of the source of energy.*

Actually, spotless flares occur more frequently than intense flares (Ruzdjak et al., 1989). Because of such a simple feature, spotless flares have been almost dismissed or forgotten these days, but they are most useful in studying the source of the energy of solar flares.

Although a photospheric dynamo was abandoned long time ago (too slow in explaining the explosive feature), it was also forgotten that an explosiveness is, in general, caused by a sudden release of *accumulated* energy. Thus, we are going to show, first of all, that a photospheric dynamo can provide the necessary power for solar flares, so that it is not necessary to consider the uncertain magnetic reconnection. We can show that the accumulation of the power for several hours is enough on the basis of observation.

Because of their simplicity, let us consider first spotless flares. They provide some crucial hints and suggestions for the cause of solar flares.

Here, as in the case of auroral substorms, *we begin with a dynamo as the power supply, a photospheric dynamo process. This is the electric current approach.* This is the major difference from the approach based on magnetic reconnection.

(b) Photospheric dynamo

The photospheric dynamo process ($V \times B$) was studied by Lee et al. (1998), Choe and Lee (1996) and Akasofu and Lee (2019). They considered: (1) An anti-parallel plasma flow along the neutral line under a magnetic arcade, which is often observed. (2) The speed of the flow is 1.5 km/s; such a flow was observed (Yang et al., 2004); a higher speed flows were observed for intense flares (cf. Min and Chao, 2002).

The important point here is that those values are *commonly observed* in the photosphere. It is difficult to understand why such an obvious power source has not recently been studied.

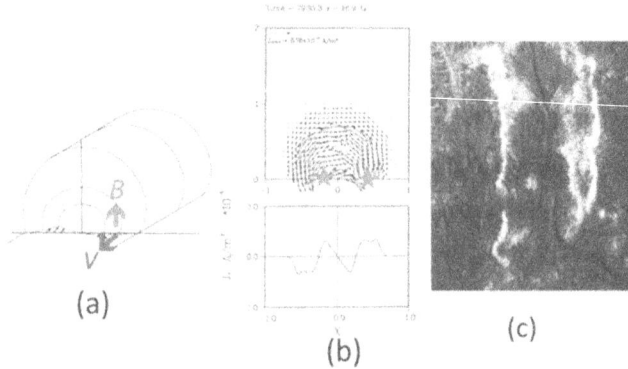

(a)

(b)

(c)

Figure 7.4 (a) The photospheric dynamo process under a magnetic arcade. (b) The field-aligned current along the arcade magnetic field lines (Courtesy of G. S. Choe); the red stars mark the locations of two flare ribbons. (c) An example of spotless flares (Svestka,1976)

The power of a dynamo is the same as that is used for aurora substorms (Section 5.2). It is given by:

$$P = \int \boldsymbol{S} \cdot d\boldsymbol{A}$$

and the Poynting vector

$$\boldsymbol{S} = (\boldsymbol{E} \times \boldsymbol{B})/4\pi$$

$$P = VzBzByA/4p,$$

where A is the photospheric surface area of the sheared magnetic arcade, (the area of typical two-ribbon of the emission). For example, we consider $A = 2$ x 2 x10^5 km x 2.5 x 10^4 km = 10^{20} cm^2 (10^{16} km^2); and $ByBz = (15$ G$)^2$ and photospheric plasma speed $V = 1.5$ x 10^5 cm/s (1.5 x 10^3 m/s), With these paramers, the power P is

$$P = 2.0 \text{ x } 10^{19} \text{ W } (= 2.0 \text{ x } 10^{26} \text{ erg/s}).$$

The above power is enough for the two-ribbon Hα emission (Svestka, 1976).

Thus, the photospheric dynamo (*based on the observed parameters*) considered here can generate enough power for two-ribbon flares (Hα) at the feet of the arcade. For intense flares, we can simply increase the speed V and the magnetic field intensity B; in particular intense flares tend to occur in an active sunspot group, where B is much higher (100 G) in active regions than that considered in this simple dynamo (15 G).

This is the reason why it is generally considered that intense solar flares are directly associated with sunspots. Figure 7.5 shows the electric currents around a spotless flare.

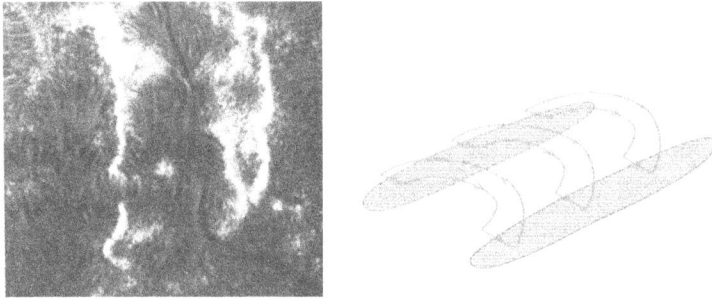

Figure 7.5 Left: Basic features of solar flares. Two ribbons are bridged by an arch-like structure (Svestka, 1976).

(c) Flare circuit

(i) Field-aligned current 1

We showed in the above that our photospheric dynamo can produce the necessary power. However, it is not enough to explain the two-ribbon emission. In order to complete the dynamo circuit, field-aligned current with the double layer is needed to ionize hydrogen atoms for the two-ribbon emission *in the chromosphere*. Thus, the current-carrying electrons have to have enough energy to ionize the chromospheric layer. We encountered a similar situation in the aurora (Section 6.4).

Our photospheric (magnetic arcade) dynamo can generate field-aligned current along the magnetic field lines of the magnetic arcade. The intensity of each field-aligned currents is 0.5×10^{-4} A/m^2 in our dynamo model (Figure

7.4). For intense solar flares, the intensity of the field-aligned current is much higher.

The Hα emission in the chromospheric layer requires electrons of 100 KeV, observed also by bremsstrahlung observations (Stix, 2002), not the ionization potential of hydrogen atom (13 eV). It is difficult to accelerate electrons to very high energy in a rarified plasma (particularly along magnetic field lines) without the double layer associated with field-aligned current (Section 6.4).

The presence of the double layer in field-aligned current is well confirmed by *in situ* satellite observations above the aurora; the aurora would not be observed without field-aligned current with the double layer; auroral current, carrying electrons of 10 KeV (not the ionization potential of 13 eV of oxygen) can penetrate into the E layer of the ionosphere (100 km in height) in order to close the magnetosphere-ionosphere current circuit (Section 6.4).

The circuit in Figure 7.4 is situated in the chromosphere, where the Hα emission is supposed to occur. The chromosphere is not a well-defined layer with spicules, and is poorly ionized (Zirin, 1988). Since the circuit is in the chromosphere, it is expected that the double layer is located in the transition region between the photosphere and the chromosphere.

Note that this situation is also the same for magnetic reconnection, which is supposed to occur in the coronal level (Figure 7.13c), unless it can generate 100 KeV electrons. It is not the problem of 13eV electrons raining down from the corona. Field-aligned current with the double layer is needed. This means that the fast plasma flow produced by magnetic reconnection has to generate field-aligned current.

(ii) Field-aligned current 2

In addition to the field-aligned currents along the magnetic arcade, our photospheric dynamo model can generate another field-aligned current system, the *loop current* system, which is considered to flow along the 'dark filament' between the two ribbons, but on the top of the magnetic arcade (Akasofu and Lee, 2019).

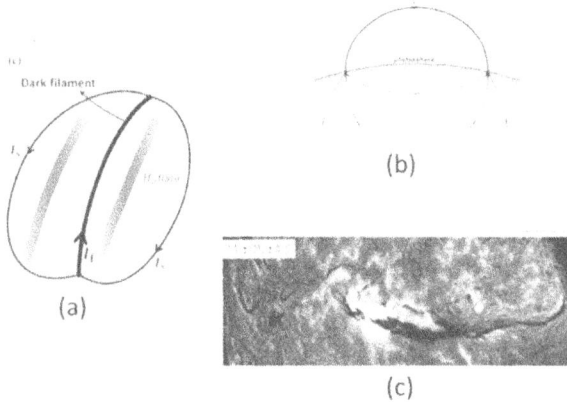

Figure 7.6 (a)The dynamo process has another current system above the two-ribbon emission between the two ribbons, but above them. It is thought that the electric current flows along the dark filament. (b) The loop current above the two ribbons (Alfven, 1950). (c) An example of dark filament. Note that a weak flare is in progress (The Norikura Solar Observatory, E. Hiei).

The loop current has magnetic energy of ($[1/2]\ I^2L$), where I and L denote the current intensity and inductance, respectively. For a typical value $I = 10^{11}$-10^{12} A (Chen and Krall, 2003) and L = 2000 H (Alfven, 1950; 1981), so that the magnetic energy of the loop is 10^{25} J (10^{32} erg). Therefore, the loop current along the dark filament can have energy for the explosive feature. Thus, we consider the loop current accumulates the needed for the explosive energy.

(d) Explosive energy release: Disparitions brusques

It is the loop current and its energy which are responsible for the sudden brightening of the two ribbons and the explosive activity of solar flares when its energy is released (unloaded).

(i) Disparitions brusques (DBs)

In supporting our idea of the loop current along the dark filament as a source of flare energy, one of the important phenomena associated with solar flares is what is called *"disparition brusques"* (Svestka,1976), but which have not been discussed much in recent years.

The dark filament between the two-ribbon (but above it) disappears at about the time of flare activity is enhanced. This is because the loop current along the dark filament explodes.

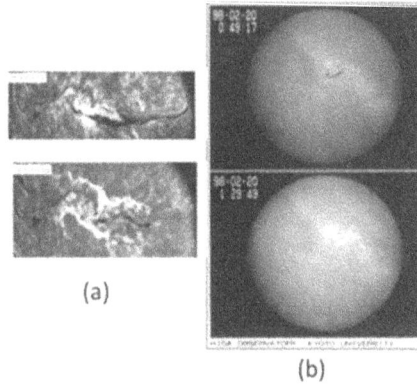

(a)

(b)

Figure 7.7 Left: (a) Upper: The dark filament prior to flare onset. Lower: Soon after flare onset, showing the disappearance of the dark filament. (b) A large-scale view of disparition bruseque. Note the disappearance of the dark filament (The Norikura Solar Observatory, E.Hiei)

(ii) Accumulation of the power

An explosive phenomenon is most often a sudden release of accumulated energy. The explosive feature of solar flares may be also such a phenomenon.

There is no observation to indicate the growth of current along the loop. However, our dynamo can cause *magnetic shear* along the center line of the magnetic arcade, which might provide an indication of the growth of the loop current. As an example, Wang et al. (1994) observed an increase of the shear angle from $40°$ to $45°$ in a rectangular area of $6.5 \times 10^8 \, km^2$ ($50° \times 25° = 3.6 \times 10^4 \, km \times 1.8 \times 10^4 \, km$) for about 5 hours before flare onset. From the magnetic shear observation, it is possible to infer the speed V of the photospheric plasma. In this case, the speed is estimated to be $V = 1.3$ km/s. This speed agrees with the flow speed, 1.6 km/s, observed by Yang et al. (2004); for the magnetic field intensity, it is assumed to be 100 G and the width of the arcade $1.8 \times 10^4 \, km$. The estimated power is 2.8×10^{19} W (2.8×10^{26} erg/s).

Thus, since the power accumulation period was about 5 hours, the accumulated energy is estimated to be 1.2×10^{25} J (1.2×10^{32} erg); in fact, an intense flare was observed in this case. Thus, the explosive feature of solar flares (corresponding to the expansion phase of auroral substorms) may be explained in this way. At the end of this chapter, it is pointed out that the distribution of photospheric current could be deduced from Figure 7.8 (left).

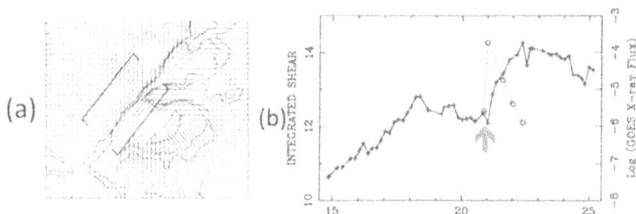

Figure 7.8 (a) Developing magnetic shear, indicating accumulation of magnetic energy. (b) Time development of the shear. Flare onset is indicated by red arrow (Wang et al.,1994). Note that the magnetic shear was even increased at flare onset (red arrow).

Incidentally, in this example, the magnetic shear was increased even at flare onset (the red arrow in Figure 7. 8), suggesting that the dynamo power is increased (instead of expending magnetic energy); this is contrary to magnetic reconnection theories, which suggest a drastic decrease of the shear and magnetic energy at flare onset.

I had an opportunity to discuss on whether or not the above observed magnetic share *before* their analysis. Harold Zirin and his group inferred that the magnetic share would decrease (because the magnetic energy is consumed by magnetic reconnection), while I suggested that the magnetic shear would not change or even increase, because the dynamo process may still be in progress. The analysis showed that magnetic shear was even increased at flare onset; this discussion was described at the end of the paper by Wang et al. (1994) in Astrophysical Journal (ApJ). *This period may correspond to the growth phase.*

A simulation of disparition brusques on the above-mentioned dynamo was successfully made by several researchers (cf. Choe and Lee, 1996). DBs are likely to be caused by a current instability in the loop current such as the kink instability (cf. Liu et all. 2007). Chen and Krall (2003) studied also the dynamics of exploring filament.

It is known that there are similar flares occur at the same location a few hours later, suggesting that the dynamo process works for many hours.

The disappeared filament on the solar disk 'reappears' as a bright and explosive prominence beyond the disk, when they become visible well above the solar disk. Indeed, Kurokawa et al. (1987) showed that the exploding prominence of a helical structure shows an unwinding feature, indicating reduction of the current and thus reduction of the loop energy.

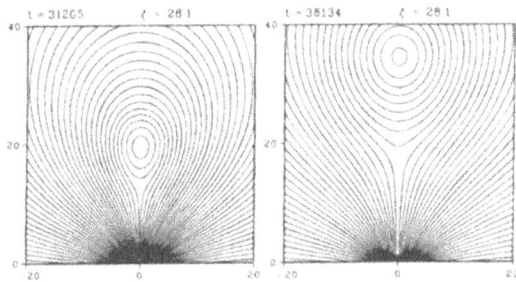

Figure 7.9 Simulation study of the magnetic field produced by the current along the dark filament. It moves upward (Choe and Lee, 19996).

Walter O. Roberts, director of the High-Altitude Observatory, University of Colorado (later, the founder of. The National Center of Atmospheric Research [NCAR]), gave me the film of an exploding prominence obtained by himself at the Climax Solar Observatory; it has been nicknamed "Grandpa" among solar physicists. One can see clearly a helical structure in the prominence; the helical structure indicates field-aligned currents along the loop.

I owe Walter greatly for stimulating me on solar physics. The High Altitude Solar Observatory of the University of Colorado was very stimulating place for me.

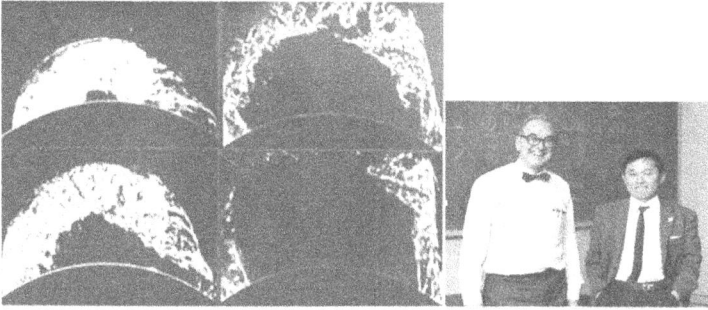

Figure 7.10 Left: A large expanding prominence, called "Grandpa" observed by Walter O. Roberts. Right: With Walter O. Roberts, the founder of both the High Altitude Observatory, University of Colorado and the National Center of Atmospheric Research (NCAR).

(e) Explosive process

In understanding solar flares, we have similar problems, such as how two *sheet* fields-aligned currents can be formed. They can be formed along the combined field lines between the magnetic arch and the magnetic field generated by the loop current in order to explain the sudden brightening of the two-ribbon emission (corresponding to the sudden brightening of two auroral arcs (in both N/S hemispheres). Thus, it is worthwhile for solar physicists and magnetospheric physicist to work together how the disrupted cross-tail current and loop current cause a sudden brightening of both auroral arcs and the two ribbons. *This explosive process might correspond to the expansion phase.*

Studies of auroral substorms faces a very similar situation, in which the cross-tail current is disrupted by its instability and increases the activities in both the northern and southern auroral oval. It is found that the magnetosphere develops an internal circuit (Bostrom's meridional current system), which has field-aligned current (Section 6.3).

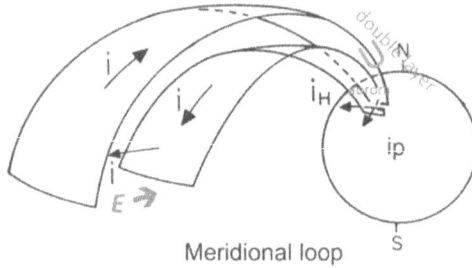

Figure 7.11 Meridional current system produced by the charge separation (***E***), originally proposed by Bostrom (1964). It is the same as Figure 6.13.

It is this field-aligned sheet current with the double layer, together with the earthward electric field (produced by magnetic field change by the disruption of the cross-tail current, $\partial Bz/\partial t)\int \partial y$), which can ionize the ionosphere enough to allow the disrupted cross-tail current to flow in the ionosphere. The sheet current is needed to explain the length of the two ribbon (2.5×10^4 km) and the thickness of the "ribbon"; see also Haerendel (2012).

It is likely that the combined field lines between the magnetic arch (Figure 7.4) and the magnetic field produced by the loop current (Figure 7.9) could make the configuration shown in Figure 7.12 (left), making the channel toward the two ribbons in order to explain the sudden brightening of the two-ribbon emission (corresponding to the sudden brightening of two auroral arcs in both N/S hemispheres).

In fact, Moore at al. (2001) showed such a combination of the two field line groups. Although it could be interpreted as magnetic reconnection, they showed that such a reconfiguration occurs *after* the two-ribbon flare is in progress (by the dynamo process in our case), not before. Note that the lower part of the Figure 7.12 corresponds to two configurations of Figure 7.11; the configuration in Figure 7.11 corresponds to one 'leg' of the loop structure in Figure 7.12, if the disrupted loop current flows along the field lines; See the middle and left of Figure 7.12a. As the loop current moves upward (Figure 7.9), it may explain why the two ribbons move part as flare progresses.

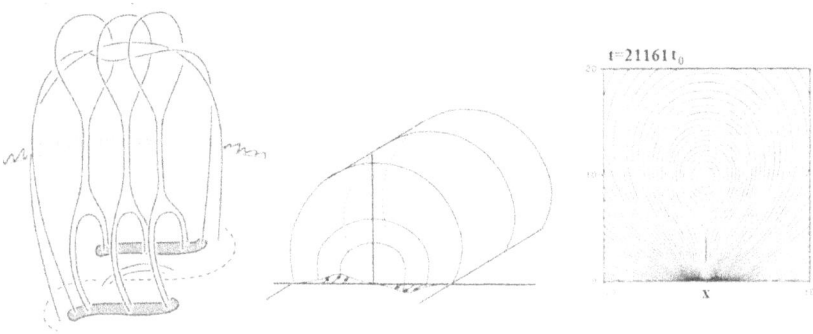

Figure 7.12a Right: Magnetic field configuration after the two-ribbon flare is in progress (Moore et al., 2001). Middle: Magnetic arcade. Right: Magnetic field of the loop current.

Figure 7.12b shows the synthesis of the above sections in a graphic form.

Solar flares

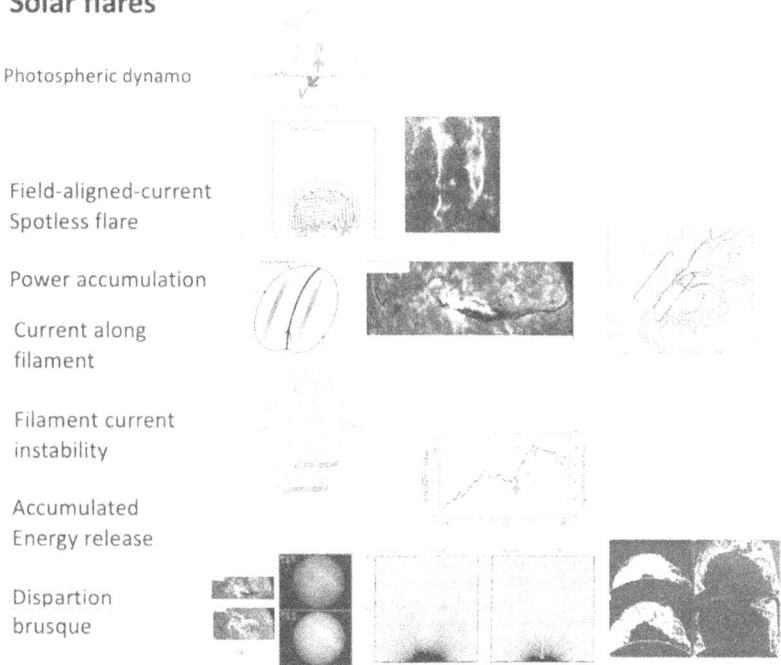

Photospheric dynamo

Field-aligned-current
Spotless flare

Power accumulation

Current along
filament

Filament current
instability

Accumulated
Energy release

Dispartion
brusque

Figure 7.12b Graphic presentation of the synthesis on solar flares.

A study of auroral substorms can provide several crucial hints based on *in situ* satellite observations of magnetic field and electric field (Akasofu et al.,

2023). In spite of a great difference of the total energy of 10^{10}, there are many similarities shown below in terms of the electric current approach.

	Auroral substorms	Solar flares
Power supply	Auroral dynamo	Photo. dynamo
Current	Cross-tail current	Loop current
Accumulation	Inner magnetosphere	Loop current
Current disruption	Cross-tail Instability	Kink instability
Mag.field change.	Dipolarization (obs.)	Not observed
Elec.field change	Observed	Not observed
Field-aligned current	Observed	Not observed
Double layer	Observed	Not observed

Since many processes in the auroral sequence were observed by *in situ* satellite observations, the whole sequence of processes requires much joint theoretical considerations as discussed in Chapter 6.

(f) Summary

Thus, our photospheric dynamo seems to explain, at least qualitatively, the major feature of solar flares, the two-ribbon emission and the explosive feature.

Thus, there is much similarity between auroral substorms and solar flares on the basis of the electric current approach, in spite of the difference of the total energy of 10^{10}.

Fortunately, each process for auroral substorms was observed by *in situ* satellite observations. However, several associated processes for solar flares have no equivalent observations. Thus, each process in the above should theoretically be examined together.

Therefore, we can learn much more about the complexity of explosive processes of auroral substorms and solar flares by learning together in different situations, rather than considering them as entirely different phenomena.

(g) Magnetic reconnection

Magnetic reconnection has long been considered, for more than 60 years, as the *only* release process of magnetic energy in an anti-parallel magnetic field configuration for solar flares and auroral substorms.

Three models of magnetic reconnection are shown in Figure 7. 13

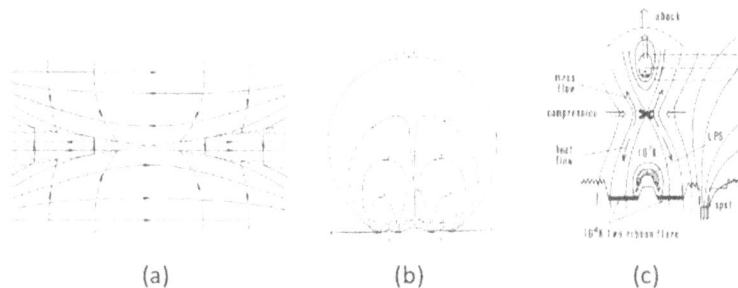

(a)   (b)   (c)

Figure 7.13 Magnetic reconnection. (a) An anti-parallel magnetic configuration and plasma flow (Vasyliunas, 1975). (b) Sweet's model (1958) of the formation of an anti-parallel magnetic field configuration by collision of sunspots. (c) A typical example of proposed magnetic reconnection in the corona (Hirayama, 1974).

Here, I present my critical review on theories of magnetic reconnection.

1. The fact that energy of 10^{25} J (10^{32} erg) is released in a short period of 0.5-2 hours during solar flares, has been one of the major issues in solar physics (requiring the energy flux of 10^{19} W (10^{26} erg/s).

2. Historically, solar physicists had eliminated photospheric dynamo as the power supply long time ago, because it was considered to be too slow for the explosive nature of solar flares. Then, they considered that magnetic energy ($B^2/8\pi$) is the source of energy for *explosive* solar flares (cf. Kiepenheuer,1953).

3. Then, the questions are:

(a)   Location

(b)   Process of releasing

4. One of the questions at that time was how the magnetic energy can be *explosively* released. Among various possibilities, Sweet (1958) proposed a model of explosive annihilation of an anti-parallel magnetic configuration, which may be produced as a result of collision of two pairs of sunspots (Figure 7.13b); see also Priest et al. (1981) and Priest and Forbes (2000).

5. However, there had and has been no *definitive* observation to support magnetic reconnection. Even if the X-line formation is observed, it is difficult to its cause-result relationship.

6. Thus, magnetic reconnection is simply a *presumption.* There is no evidence that an anti-parallel magnetic configuration '*annihilates*' itself, supplying the total energy of 10^{25} J with the rate of 10^{19} W for solar flares.

7. Actually, in as early as 1963, Parker (1963) doubted the physics of reconnection (based on physical parameters) and concluded: "The observational and theoretical difficulties with the hypothesis of magnetic field annihilation suggest that other alternatives for the flares must be explored". Alfven (1986, p.784) called magnetic reconnection (merging) as "pseudoscience".

8. However, many theorists were interested only in the presumed annihilation process, not solar flare or substorm phenomena. Indeed, it was not their concern how the generated fast plasma flow from magnetic reconnection can account for various flare or substorm processes, such as the two-ribbon Hα emission (a major energy dissipation requiring 100 KeV electrons). They tried to show that magnetic reconnection is explosive.

9. Meanwhile, theorists are so convinced about the validity of the concept of magnetic reconnection, many observers reported observed X-line formation *without confirming, whether it is the cause of solar flares,* just supporting magnetic reconnection. They were interested only in physics of magnetic

212

reconnection. It is for this reason Hudson and Khan (1996) were very critical about such a support of observations. The same problem exists for auroral substorms; see also Fletcher et all (2011).

10. There are many cases when solar flares are not associated with the X-line formation in the corona. Moore at al. (2001) examined six major solar flares and concluded: "All six events are single-bipole event in that during the onset and early development of the explosion they show no evidence for reconnection between the exploding bipole and any surrounding magnetic fields." In fact, Moore et al. (2001) noted: "the reconnection as a byproduct."; in such a case, the term "reconnection" should be "linkage", because magnetic reconnection is not the power generation process.

11. Many observations do not seem to conform with magnetic reconnection were dismissed. Figure 7.14 shows an example, in which the formation of the X-line above a solar flare is not evident (compared with Figure 7.13c.

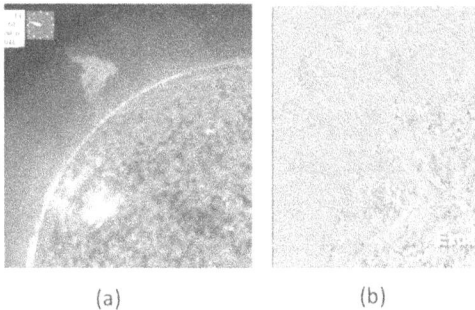

(a)   (b)

Figure 7.14 (a,b) Example of flare without the X-line formation (without magnetic reconnection) in the corona; it shows an excellent image of magnetic field lines above the flare (Courtesy of A. Title).

There are many other examples, in which the X-line formation in the corona is not obvious. Figure 7.15 is an example. Compare it with Figure 7.13c (unless magnetic reconnection occurs in the chromosphere).

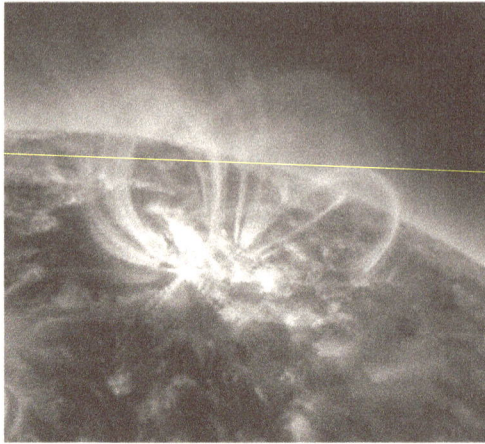

Figure 7.15 An intense flare with no clear presence of the X-line between two loop fields (K. Shibata). It is interesting to compare this with Figure 7.13c by Hirayama (1974).

14. Thus, the observed "reconnection" is likely to be linking of magnetic field lines without energy production, just like at the front of the magnetopause, where IMF field lines and magnetospheric field lines link (providing the ground for auroral dynamo (Section 5.2, Figure 5.5).

15. Meanwhile, magnetospheric physicists discovered nearly anti-parallel magnetic configuration in the magnetotail (Ness, 1965). In the year before, the concept of auroral substorms was published (Akasofu, 1964), so that many magnetospheric physicists have jointed in studying magnetic reconnection as the source process of auroral substorms by sending many satellites in the magnetotail (cf. Angelopoulous et al., 2008; Burch et al., 2016). It may be noted that the above criticisms on solar flares on magnetic reconnection are also applicable for auroral substorms.

16. Actually, magnetospheric physicists became more enthusiastic than solar physicists on magnetic reconnection, stating that they are the only one who can prove *in situ* the most important energy source of many astrophysical, solar and magnetospheric phenomena. Thus, a number of satellites were launched into the magnetotail by several countries in observing and determining parameters related to magnetic reconnection without obtaining any parameter in determining the reconnection rate (Burch et al., 2016).

17. Recently, Torbert et al. (2018) studied results from the MMS mission. Reviewing their studies and others, Hesse and Cassak (2020) elaborated further electron diffusion in the reconnection region, but found difficult to examine magnetic reconnection in the magnetotail for auroral substorms; they stated that magnetic reconnection is "illusive" and "mysterious". Thus, they examined mainly magnetic reconnection (linkage) on the magnetopause, *where no explosive phenomenon reported* (further, *no references on solar flares).* In fact, they consider that Parker's criticism is in some sense (in terms of physical parameters) still valid.

After all, an important lesson is: *In natural sciences, unlike basic physics, it is important that a solid observed fact should lead theoretical considerations, not the other way around;* this point is discussed more in Chapter 8.

My motive

Looking back, my study of solar flares began when I found *spotless flares* in the book by Svestka (1976). I was struck by their simplicity. Before that time, I read many papers on solar flares, in which flares occurred among active and complex sunspot groups. Obviously, they were intense flares (which have many features to report), but it seemed that spotless flares were forgotten.

Thus, spotless flares are an exceptional or even 'odd' case among all other intense flares. I found that spotless flares have the two-ribbon emission, which has one of the largest total energy dissipations among other flare features (like the production of sub-cosmic ray has much less total energy). Thus, it was obvious to me that we can learn the basic aspect of solar flares by studying simplest spotless flares.

From the 1960s, magnetic reconnection has been overwhelmingly popular as the source of flare energy. However, they rely on the explosive feature of solar flares to presumed magnetic reconnection, which is still very uncertain.

Thus, I thought that I should consider a much simpler process of photospheric dynamo, which is *verifiable by photospheric observations*. Thus, I ask Lou-Chuang Lee to join me when he was a professor of the Geophysical Institute of the University of Alaska Fairbanks and has continued to work together afterward at the Institute of Earth Science, Academia Sinica in Taiwan (Akasofu and Lee, 2019).

One important point here is that spotless flares provided me the opportunity to study solar flares, rather than many recent interesting features of solar flares. Thus, going back to old data has been useful in considering a new way of studying solar flares. Asking for better special and time resolution data and looking for new observational facts are certainly desirable, but studying old subjects is very important as well. In some cases, we have to go back to the problem of energy source.

(h) Coronal mass ejections (CMEs/MCs)

As mentioned in (e), the dark filament erupts. The ejected filament (or solar gas) is called coronal mass ejection (CME) or magnetic cloud (MC). However, their topology (both geometric and magnetic, as well as magnetically attached to or detached from the sun) is not yet very clear, in spite of a large number of theoretical and observational papers; see Bulraga (1995), Marubashi (1997), Lugas and Roussev (2011), Xu et al. (2020), Lepping et al.(2003), Janvier et al. (2013, 20214), Zhang et al. (2013) and many others.

Figure 7.16 Example of CMEs/MCs leaving the sun, together with a shock wave (NASA).

It has been considered by some researchers that some of CMEs/MCs are rooted on the photosphere. An important point here is that *electric currents* flow along its expanding magnetic loop from the sun, because they are likely to be an expansion of *disparition brusques (DBs) [a, i)]*; see also Chen and Krall (2003).

(i) Propagation of solar flare effects to 1 au

One can infer how the *energy* of CMEs spreads in interplanetary space from the sun based on the relationship between the central meridian distance of flares and the intensity of geomagnetic storms measured by the geomagnetic indices.

Using the intensity of the main phase decrease (as we learned in Chapter 2) recorded at Honolulu, it was found that the flare effect *in terms of energy flow* is confined in a rather narrow space of about 60° in longitude from the sun, although the shock waves expand into interplanetary space much widely. The average transit time of coronal mass ejections (CMEs) to the earth (1 au) is about 40 hours.

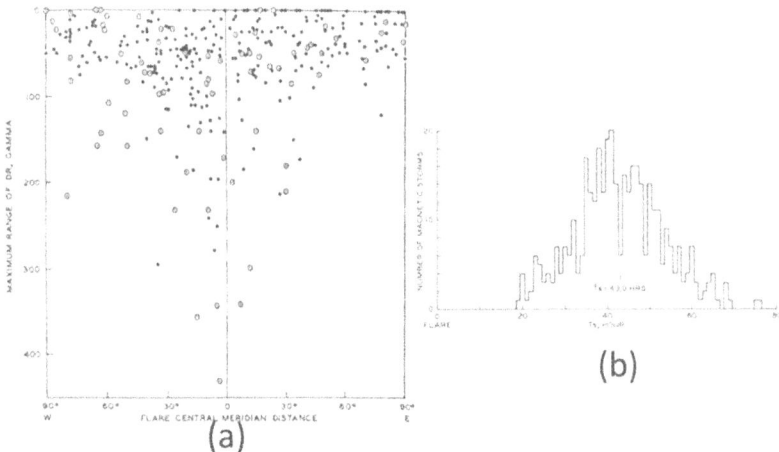

Figure 7.40 (a)The dependence of the intensity of geomagnetic storms on the central meridian distance of solar flares (geomagnetic storm record at Honolulu). The dot with a circle is intense flares which produced solar sub-cosmic rays. (b) The transit time of CMEs

(shock waves); Yoshida and Akasofu (1965).

In order to develop a simple simulation method (instead of an extensive MDH simulation scheme), we developed our own simulation scheme called HAF scheme (Hakamada and Akasofu, 1982; Akasofu and Fry, 1986b).

First of all, we developed the propagation of the solar wind when there is no solar flare activity.

The basic concept of the HAF scheme is briefly given here. The wind distribution is a sinusoidal on the photosphere; the reason of the sinusoidal distribution, see Section 7.3 (c) for details.

As the sun rotates, *a fixed point in space* (not fixed on the sun near the photospheric surface) observes two half-sinusoidal variations of the speed as a function of time during one rotation of the sun. The amplitude of the sinusoidal pattern can be changed, depending on solar conditions during the sunspot cycle by knowing the magnetic equator on the source surface (a concentric spherical surface of three solar radii); see Section 7.3 (c).

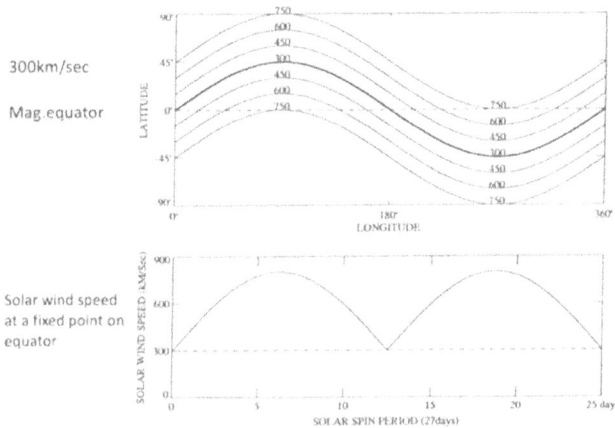

Figure 7.17 Upper: Speed distribution of the solar wind on the photosphere (Carrington map). Note that the magnetic equator (the speed is 300 km/s) is inclined from the ecliptic equatorial plane. Lower: The speed variation during one rotation of the sun at a fixed point near the photospheric surface, but not fixed on the sun.

As the two half-sinusoidal waves propagate outward, the speed of each wave

218

as a function of distance from the sun can be simulated by an ordinary MHD simulation (cf. Dryer, 1994); a distinct shock wave structure is formed at a distance of about 7 au. Our scheme follows this propagation result.

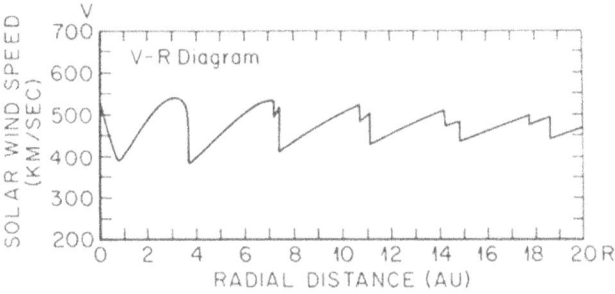

Figure 7.18a The resulting speed distribution of one of the half-sine wave as a function of distance from the sun. Note that a distinct shock is formed at about 7 au.

As a test, the pattern of the interplanetary magnetic field (IMF) within 2 au for the above conditions is calculated. It shows Parker spiral structure. Because of the non-uniform speed distribution, two shock wave-like structures are generated, so that there occur two co-rotating structures. We computed also the magnetic equatorial plane to 2 au based on the magnetic equator on the source surface (CR 1654).

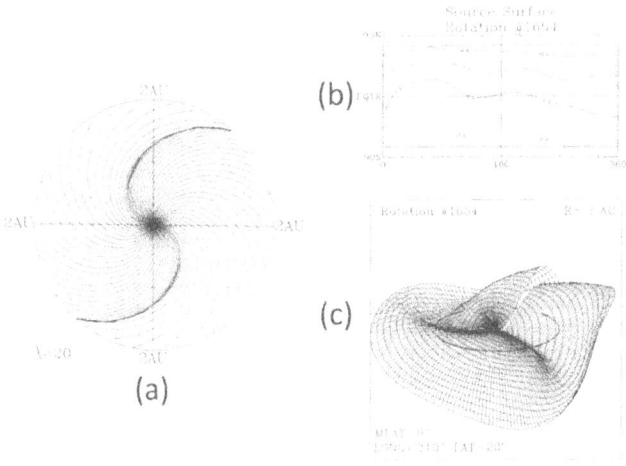

Figure 7.18b (a) The magnetic field on the equatorial plane. (b) The magnetic equator (00) on the Carrington map (CR 1654). (c) Corresponding magnetic equatorial plane.

After a quiet situation is learned, a solar flare is represented by adding a circular speed distribution at the reported location of flares (the longitude, latitude) in the Carrington map. The intensity of flares can be adjusted in terms of the radius of the circle and the speed at the center of the circle. When the sun is active, many flares occur. They can be added in the same way on the same day, next day or after weeks or months later (Akasofu and Fry, 1985, 1986b). The accuracy of this method was compared with one of MDH simulation methods in terms of the arrival time (only thing to be able to compare). It was found that the result (accuracy) is about the same (Sun et al., 1985). This is because there are many uncertainties in choosing even the initial conditions, how CMEs are ejected (disparition bruseques, Section 7.1 [a]).

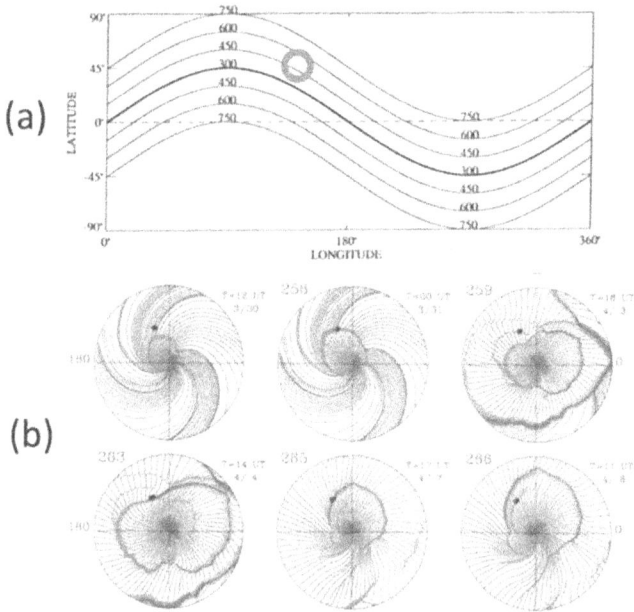

Figure 7.19 Simulation of a successive flares. Upper: An example of the reported location of the flare is shown by a circle on the Carrington map. The intensity of flare is given in terms of the size of the circle and the speed at its center. The simulation is made by taking into account the successive location of flares relative to the earth. Lower: An example of simulation in the case of successive flares. The red lines indicate positive (away from the sun), and blue negative (toward the sun).

Saito et al. (2007) attempted to construct the spiral magnetic field structure of

CMEs/MCs by assuming that an expanding current loop from the sun has electric currents of 10^9 A at the distance of the earth (initially 10^{11} A); they tried to determine the helical structure of the magnetic field of CME/MC by trials and errors in such a way that the helical structure thus constructed agrees with the IMF changes during its passage at the earth (see Figure 7.20 [b] black [observed] and red [simulation]); it was a very humble effort as shown in the figure below.

This is one of the earliest attempts to construct the magnetic configuration of CME/MC *on the basis of observations by assuming electric current from the sun (not on MHD theories)*.

Such a cumbersome effort is generally avoided (often easily be criticized), and this is why the prediction by Space Weather groups cannot progress much. It is this kind of efforts, which are needed for the prediction of IMF (-Bz) or even the Kp index (or even the largest Kp) as a function of time.

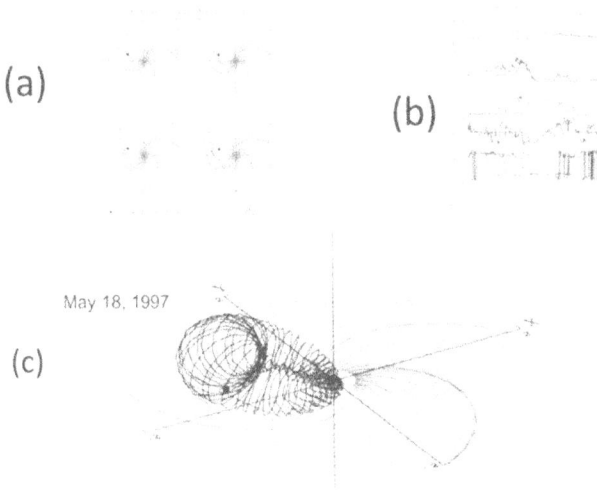

Figure 7.20 (a)The simulated propagation of shock wave on May 18, 1997 event (red, IMF outward, blue inward); the location of the earth is indicated by a dot. (b) The comparison of the observed (black) and simulated (red) changes of the solar wind. (c) An example of helical structure which agrees with the observed (black) magnetic changes at the earth in (b); the location of the earth is indicated by a dot.

There is one way to detect interplanetary shock waves (or CMEs/MCs) between the sun and the earth by observing the scintillation of radio stars. Hewish et al. (1985) projected the observed scintillation in a sky map. The simulated shock wave for the same responsible flare was reproduced in the sky map (Akasofu and Lee, 1989). There is a reasonable agreement between them, so that it may be possible to detect the shock waves by combining the scintillation observations and the simulation studies.

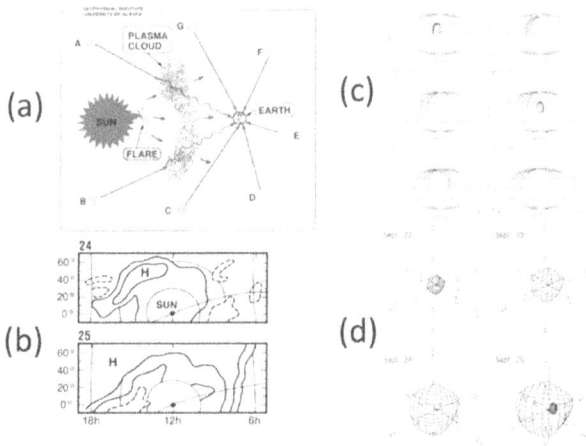

Figure 7.21 Schematic illustration, showing the detection of a shock wave in the sky on the basis of radio star scintillation. (b) The sky map of observed scintillation of radio stars (Hewish et al. [1985]). (c) The projection (sky map) of the expanding shock wave simulated for the event in (a). (d) A 3-D view of the shock wave (Akasofu and Lee,1989).

(j) Space weather

The space physics community has established a "space weather" research group. The term "weather" suggests that they will be able to forecast at least the intensity of geomagnetic storms. As mentioned in the above, the only sure way to predict the intensity of geomagnetic storms is to be able to predict the intensity of the IMF [-Bz] component as a function of time. Thus, the success of Space Weather Forecasting depends on quantitatively forecasting the magnetic topology of CMEs/MCs.

For each observed solar flare/geomagnetic storm, they should form a

222

dedicated group, which work together on the following:

1. Study of exploding the dark filaments (disparitions brusques) by solar physicists.

2. Study of the magnetic configuration of CMEs/MCs by interplanetary physicist. s.

3. Study of geomagnetic storms in terms of IMF (-Bz) by magnetospheric physicists.

At the present time, solar physicists are not really concerned about the IMF Bz, while magnetospheric physicists consider that they are interested in IMF (-Bz) only when it arrives at the front of the magnetosphere.

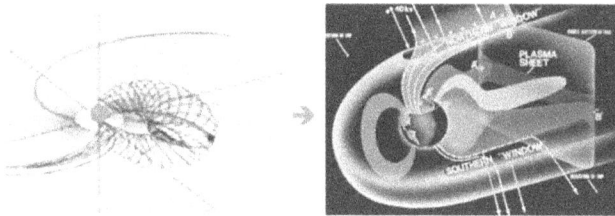

Figure 7.22 Interaction between CME/MC with magnetosphere. We have to understand the interaction between CME magnetic field and the magnetosphere for storm prediction.

In Section 7.3(d), we succeeded in reproducing recurrent geomagnetic storms as a function of time on the basis of the magnetic equator on the source surface of the sun, so that it is possible now to predict them on the basis of solar data.

Therefore, it is now possible to predict the development of 27- day recurrent geomagnetic storms as a function of time on the basis of the solar condition.

(k) Heliospheric disturbances

(i) Heliospheric disturbances to 2 au

We can extend the method in (i) beyond 1 au. In April, 1973, three shock waves were observed by two space probes, HELIOS A, B and the earth. The inferred shocks by the observation (Bulraga et al. (1981) can fairly well be reproduced by our simple simulation. The agreement suggests that both inferred and simulated results support each other.

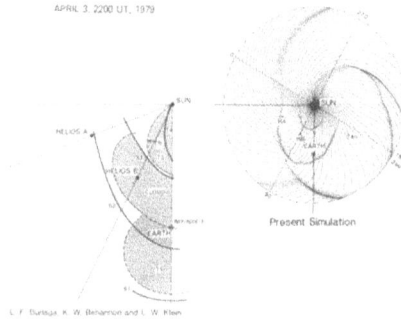

Figure 7.23 Example of multi-shock wave event during the end of March to the beginning of April of 1979 (Bulraga et al., 1981) and its simulation result. Three shock waves are reasonably well reproduced. Note that the simulation provided the shock structure in the whole area of 2 au (Akasofu, 1996).

Next, we show our simulation of shock waves up to 10 au, when both Ulysses and Cassini were operating.

(ii) Heliospheric disturbances to a distance of 10 - 30 au

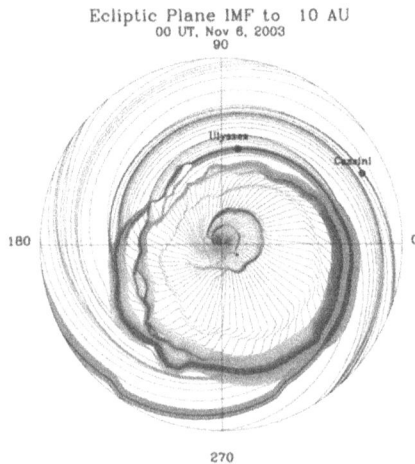

Figure 7.42 Several propagating interplanetary shock waves up to 10 au. when Ulysses and Cassini were operating.

Then, interplanetary disturbances are simulated within a distance of 30 au, and the results are also compared with the Pioneer 11 observation of the speed and the simulation. In spite of the simple HAF scheme, the observation of speed at Pioneer 11 and the simulated result agrees reasonably well. During the same period, the sun was very active and several shock waves were propagating in all directions.

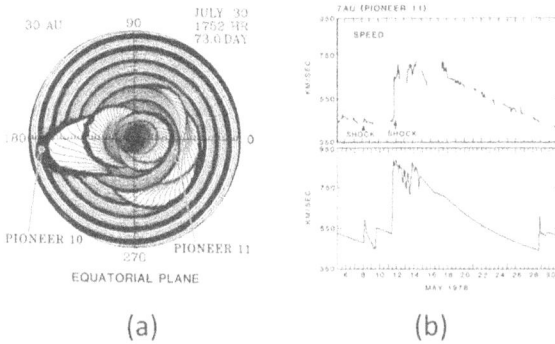

(a)   (b)

Figure 7.24 (a) Simulation of interplanetary disturbances to a distance of 30 au; the location of Pioneer 10 and 11 is indicated. (b) Comparison of the observed and simulated disturbances at the location of Pioneer 11 (Akasofu et al. (1985)

(iii) Heliospheric disturbances to 100 au

The HAF scheme is extended to examine disturbances within a distance of 100 au, which includes about 200 days of events in an early 2004. All the events during about 200 days prior to February 16, 2004 are superposed. Shock waves tend to merge together at great distances, as later shock waves with faster speeds catch up with the earlier ones, forming a "magnetic barrier." It can also be seen that after the disturbed period, a quiet condition had resumed as shown by a steady spiral structure in the inner interplanetary space.

It is hoped the HAF or more improved simulations will be used for future deep space prove observations, because the method is simple enough to infer heliospheric conditions in studying the data on a daily basis as demonstrated, instead of elaborate MHD simulations.

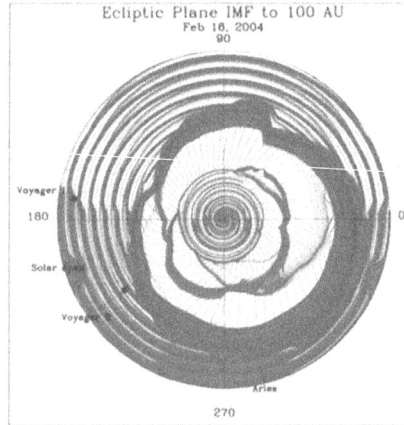

Figure 7.25 Simulation of heliospheric disturbances on February 16 in 2004 up to 100 au; all the events prior that date are superposed. All the shock waves produced a barrier, which may prevent the entry of cosmic-rays to penetrate into the inner part of the heliosphere and might be useful in explaining the 11-year variation of the intensity of cosmic-rays, when cosmic rays arrive from the outer surface of the heliosphere.

Note on electric current distribution

As mentioned earlier, it is not possible to measure directly electric current in space or the ionosphere. In magnetospheric physics, we determine the current on the basis of the ground-based magnetic field observation (the horizontal component, H, D), as shown in Figure 5.11; its (c) is reproduced here on the left side. The deduced electric current distribution is also reproduced.

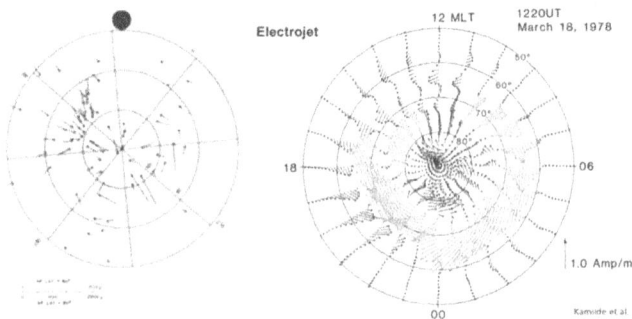

The magnetic field on the photosphere (the horizontal component) can be measured as Figure 7.8 shows.

There must be a way to deduce the electric current distribution, as has been done in magnetospheric physics (Section 5.5). Its *div I* could also determine the field-aligned current.

Both magnetospheric physics and solar physics have to rely on the magnetic field measurement in determining the electric current distribution.

References

Introduction

Aschwanden, M., 2005, *Physics of the Solar Corona*, Springer, in association with Praxis Pub., Chichester, UK.

Babcock, H. W., 1961, The topology of the sun's magnetic field and the 22-year cycle, ApJ., **133**, 572.

Fletcher, L. et al., 2011, An observational overview of solar flares, *Space Sci. Rev.*,**159**, 19. https://doi.org/10.1007/s11214-010-9701-8

Hesse, M. and Cassak, P. A., 2020, Magnetic reconnection in space sciences: past, present and future, J. Geophys. Space Phys. Res.,**125**, e2018JA025935. hattps.//doi,org/10.1029/2018JA025935

Viall, N. and Borovsky, J. E., 2020, Nine outstanding questions of solar wind questions, J. Geophys. Res. Space Physics, **125**, e2018JA026005.

Solar flares

Akasofu, S.-I., 1964, The development of the auroral substorm, Plant. Space Sci., **12**, 273.

Akasofu, S.-I., 1996, New scheme provides a first step toward geomagnetic storm prediction, *Eos, 77*, 225.

Akasofu, S.-I., Hakamada, K. and Fry, C. D., 1983, Solar wind disturbances caused by solar flares: Equatorial plane, Planet. Space Sci., 31, 1435.

Akasofu, S.-I., Fillius, W., Sun, W. Fry, C. D. and M. Dryer, M., 1985, A simulation study of two major events in the heliosphere during the present sunspot cycle, J. Geophys. Res., **90**, 8193.

Akasofu, S.-I. and Fry, C. D., 1985, Heliospheric current sheet and its solar cycle variation, J. Geophys. Res., **91**, 13679.

Akasofu, S.-I. and Fry, C. D., 1986a, A first generation numerical geomagnetic storm prediction scheme, Planet. Space Sci., **34**, 72.

Akasofu, S.-I. and Fry, C. D., 1986b, Heliospheric current sheet and its solar cycle variations, J. Geophys. Res., **91**, 13679.

Akasofu, S.-I. and Fry, C. D., 1986b, Heliospheric current sheet and its solar cycle variations, J. Geophys. Res., **91**, 13679.

Akasofu, S.-I. and Lee, L.-C., 1989, Modeling of an interplanetary disturbance event tracked by the interplanetary scintillation method, Planet. Space Sci., **37**, 73.

Akasofu, S.-I., Watanabe, H. and Saito, T., 2005, A new morphology of solar activity and recurrent geomagnetic disturbances: The late-declining phase of the sunspot cycle, Space Sci. Rev., **120**: 27, Doi: 10. 1007/s1124-005-8052-3.

Akasofu, S.-I., 2023, A new understanding of why the aurora has explosive characteristics, Mon. Not. Roy. Astronom. Soc., **518**, 3286, https://doi.org/10.1093/mnras/stac3187

Alfven, H, 1950, *Cosmical Electrodynamics*, Oxford Univ. Prress

Alfven, H., 1968, The second approach to cosmical electrodynamics, in *The Birkeland Symposium on Aurora and Magnetic storm,* ed. by A. Egeland and J. Holt, Centre National de la Recherche Scientifique, Paris, p. 439-444.

Alfven, H., 1977, Electric currents in cosmic plasmas, Rev. Geophys. Space Phys. **15**, 272.

Alfven, H., 1981, *Cosmic Plasma*, D. Reidel Pub. Co., Dordecht, Holland.

Angelopoulos, V. et al., 2008, Tailreconnection triggering substorm onset, Science **321** (589), 931.

Burch, J. L. et al., 2016, Electron-scale measurements of magnetic reconnection in space, Science, **352**. 6290. First in situ observation of a magnetic reconnection including the crescent distribution.

Burlaga, L. F., Sittler, E. Mariani and Schwenn, R., 1981, Magnetic loop behind an interplanetary shock: Voyager, Heilo and IMP 8 observations, J. Geophys. Res., **86**, 6673.

Canfield, R. C., Hudson, H. S. and McKenzie, D. El., 1999, Sigmoidal morphology and eruptive solar activity, Geophys. Res. Lett., **26**, 627.

Chen, G. S. and Krall, J., 2003, Accerelration of coronal mass ejection, J. Geophys. Res., **108**, httpss.//doi.org/10.1029/2003JA009849.bSSH.2.

Choe, G. S. and Lee, L.-C., 1996, Evolution of solar magnetic arcades. I. Ideal MHD evolution under foot point shearing, ApJ., **472**: 360

Dodson, H. and Hedeman, E. R., 1970, Major H flares in centers of activity with very little or no spots, Solar Physics, **13**, 401.

Fletcher, L. et al., 2011, An observational overview of solar flares, Space Sci. Rev., **159**, 19. https://doi.org/10.1007/s11214-010-9701-8

Dryer, M.,1994, Interplanetary studies: Propagation of disturbances between sun and the magnetosphere, 1994, Space Sci. Rev., **67**, 363.

Fry, C. D. et at., 1985, The three-dimensional geometry of the heliospheric current sheet, Planet. Space Sci., **33**, 915.

Haerendel, G., 2012, Solar auroras, ApJ, **749**:166 (13pp),2012 April 20.

Hakamada, K. and Akasofu, S.-I., 1982, Simulation of three-dimensional solar wind disturbances ad resulting geomagnetic storms, Space Sci. Rev., **31**, 3.

Hesse, M. and Cassak, P. A., 2020, Magnetic reconnection in space sciences: past, present and future, J. Geophys. Space Phys. Res., **125**, e2018JA025935. hattps.//doi,org/10.1029/2018JA025935

Hewish, A. and Duffett-Smith, P. J., 1987, A new method of forecasting geomagnetic activity and proton shower, Planet. Space Sci., **35**, 48

Hirayama, T.,1974, Theoretical model of flares and prominences, Solar Phys. **34**, 323.

Hudson, H. S. and Khan, J. I., 1996, Observational problems for flare models based on large-scale magnetic re connection, 135, in Magnetic Reconnection in space and laboratory plasmas, ed. by R. D. Bentley and J. R. Mariska, ASP Confernce Series Volume **111**.

Janvier et al., 2013, Global axis shape of magnetic clouds deduced from the distribution of their local axisorientation, ApJ., **A50**, doi: 10. 1051/0004.6361/20132442.

Janvier, M., Demoulin, P. and Dasso, S., 2014, In situ properties of small and large flux ropes, J. Geophys. Res., 19, 7088, doi: 10. 1002/2014JA020218.

Lepping, R. P. et al., 2003, Profile of an average magnetic cloud at 1 au for the quiet solar phase: Wind observations, Solar Phys., **212**, 425-444.

Lugas, N. and Roussev, H., 2011, Numerical modeling of interplanetary coronal mass ejections and comparison with heliospheric images, JATO, **73**, 1187-1200.

Kiepenheuer, K. O., 1953, *The Sun, Solar activity*, 322, ed. by G. P. Kuiper, University of Chicago Press.

Kurokawa, H. et al., 1987, Rotating eruption of untwisting filament triggered by the #B flare of 25 April, Solar Phys., **108**, 251.

Lee, L. C., Choe, G. S. and Akasofu, S. I., 1995, A simulation study of the formation of solar prominences In: Solar plasma coupling between small and medium scale processes, **29**, AGU, Washington, DC.

Liu, Rui, Alexsander, D. and Gilbert, H. R., 2007, Kink-induced catastrophe in a coronal eruption. ApJ, **661**,1260.

Marubashi, K., 1981, Interplanetary magnetic flux ropes observed by the pioneer Venus orbiter, Adv. Pace Res. **11**, 57.

Min, S. and Chao, J., The rotating sunspot in Ar 10930, Solar Phys. **258**, 203.

Moore, R. I., Sterling, A. C., Hudson, H. and Lemen, J. R., 2001, ApJ**, 552**, 839.

Ness, N. F., 1965, The earth's magnetic tail, J. Geophys. Res., **70**, (13), 2989.

Parker, E. N., 1963, The solar-flare phenomenon. and the theory of reconnection and annihilation of magnetic fields, ApJ Suplement, **vol. 8**, 1777.

Petschek, H. E., 1964, Magnetic field annihilation, 425, in The Physics of Solar Flares, Proceedings of the AAS-NASA symposium held 28-30 October 1963 at the Goddard Space Fight Center, Greenbelt, MD, ed. by W. H.. Hess, Washington, Dc.

Priest, E. R. (ed), 1981, *Solar Flare Magnethydrodynamics*: The fluid mechanics of astrophysics and geophysics, Volume 1, Gordon and Breach Sci. Pub., New York.

Priest, E. R. and Forbes, T., 2000, *Magnetic Reconnection*, Cambridge University Press, Cambridge.

Ruzdjak, V., Vrsnak, B., Schrool A. and Brajsa, B., 1989, A comparison of Halpha and soft X-ray characteristics of spotless and spot group flares, Solar Phys., **123**, 309.

Petschek, H. E., 1964, Magnetic field annihilation, 425, in The Physics of Solar Flares, Proceedings of the AAS-NASA symposium held 28-30 October 1963 at the Goddard Space Fight Center, Greenbelt, MD, ed. by W. H. Hess, Washington, Dc.

Priest, E. R. (ed), 1981, *Solar Flare Magnethydrodynamics*: The fluid mechanics of astrophysics and geophysics, Volume 1, Gordon and Breach Sci. Pub., New York.

Priest, E. R. and Forbes, T., 2000, *Magnetic Reconnection*, Cambridge University Press, Cambridge.

Ruzdjak, V., Vrsnak, B., Schrool A. and Brajsa, B., 1989, A comparison of Halpha and soft X-ray characteristics of spotless and spot group flares, Solar Phys., **123**, 309.

Saito, T., Oki, T. and Akasofu, S.-I., 1989, The sunspot cycle variations of the neutral line on the source surface, J. Geophys. Res., **94**, 5453.

Saito, T. et al., 2007, Trans equatorial magnetic flux loops on the sun as a possible new source of geomagnetic storms, J. Geophys. Res., **112**, A05102, doi: 10. 1029/2006JA011941

Sun, W. et al., 1985, Calibration of the kinematic method of studying solar wind disturbances on the basis of a one-dimensional MHD solution and a simulation study of the heliosphere disturbances between 22 November and 6 December, 1977, Planet. Space Sci., **33**, 933.

Yoshida, S. and Akasofu, S.-I., 1965, A study of the propagation of solar particles in interplanetary space: The center-limb effect of the magnitude of cosmic ray storms and of geomagnetic storms, Planet. Space Sci., **13**, 435, 1965

Stix, M., 2002, *The Sun,* 2nd ed. Springer, Verlag, Berlin.

Svestka, Z., 1976, *Solar Flares*, D. Reidel Pub. Co., Dordrecht-Holand.

7.2 Solar corona

There have been 40 theories on the ionization of the corona (Aschwanden, 2005), but none of them considers a direct ionization by current-carrying electrons. In this chapter, we introduce the electric current approach to solve partially this difficult problem.

(a) Introduction

It was in the 1940s when coronal lights are found to be emitted from high ionized atoms, such as Fe^{XII}, which lost 14 electrons out of 26; their presence is thought to be caused by a very high temperature of about one million degrees. Thus, from the time of discovery of the ionization of Fe^{XII}, researchers have tried to explain how the heat source of the photosphere (including Alfven waves) can cause the high temperature and the ionization of the corona.

On the other hand, it is important to recognize that the corona has a loop structure, which is very bright, suggesting the ionization along the loops.

Therefore, there is another *entirely different* ionization process, the ionization caused by energetic *current-carrying electrons* along magnetic field lines with the double layer, *the electric current approach*. Actually, the ionization in the corona requires much higher energy than the ionization potential, as explained in the aurora and solar flares and auroral substorms, 100 KeV for solar flares and 10 keV for auroral substorms, instead of the ionization potential (13 eV). For this purpose, the importance of field-aligned current with the double layer was mentioned earlier in our study of auroral substorms and solar flares (Section 6.4). This is the case for. our theory of coronal ionization, because the ionizing electrons have to penetrate into the chromosphere; theories based on magnetic reconnection have the same problem.

There are at least three requirements in considering the ionization of the corona in terms of current-carrying energetic electrons.

(i) Dynamo process

(ii) Current circuit

(iii) The presence of the double layer along field-aligned current.

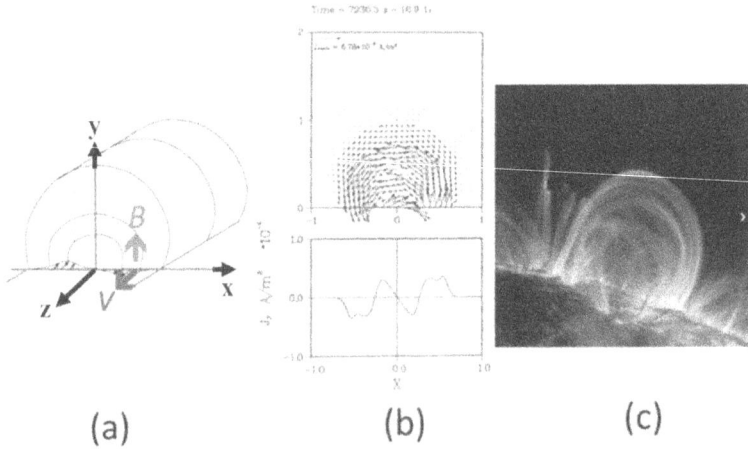

Figure 7.26 (a) A model of photospheric dynamo (Choe and Lee, 1996). (b) The field-aligned currents resulting from the dynamo along the magnetic arcade configuration in (a) [Courtesy of G. S. Choe]; this dynamo can produce field-aligned current of 50 μA /m². (c) Magnetic loops in the corona (NASA Corona Collection).

(b) Dynamo and circuit

As the source of the field-aligned currents, a photospheric dynamo process by Akasofu and Lee (2019) can generate field-aligned currents (Section 7.1).

The emission from Fe^{Xii} and other highly ionized atoms are brightest in the coronal loops (Figures 7.26 (right) and 7.27), so that the magnetic field lines must be a part of the circuit. In fact, it has been considered that the corona is not a stratified atmosphere, but mainly consisting of large number of loops (Aschwanden, 2005).

Figure 7.27 The corona seems to consist of a large number of loops, instead of a unified atmospheric layer (NASA Collection).

234

(c) Coronal ionization by field-aligned currents

The equation for the ionization rate q by a beam of energetic electrons in the ionosphere is given by Rees (1989):

$$q = F\varepsilon_e\rho/(R(\varepsilon_e^2) \times 30ev).$$

For the auroral ionization, taking electron flux $(F = 6.2$ x $10^8/cm^2$ s, corresponding to 1μ A), electron energy ($\varepsilon_e = 10$ Kev), atmospheric density (N= $10^{12}/cm^3$), mass density ($\rho = 1.6$ x $10^{-12}g/cm^3$ (= $10^{12}/cm^3$ x 1.6 x10^{-24} g)) and effective range $R(\varepsilon_e^2) = 2.5$ x 10^{-4} g/cm^2),the ionization rate is $q = 1.3$ x $10^4/cm^3$s, which is an acceptable value.

In the above estimate, we chose a low level of corona or the transition from the corona to the chromosphere (N= $10^{12}/cm^3$). The energetic current-carrying electrons thus produce the ionization along the magnetic field lines by mirroring back and forth until they exhaust the energy. Thus, it seems that the coronal ionization by current-carrying energetic electrons might be included in the future.

Further, the above ionization rate q can barely supply the solar wind. If the number of the neutral density is less than N = $10^{12}/cm^3$), it is not possible to supply the solar wind, indicating that the double layer is not located in a higher level of the corona.

Furthermore, the fact that the ionization of Fe^{XII} can partially be caused by energetic current-carrying electrons (not by the heating) suggests that the coronal temperature estimated by the ionization potential of Fe^{XII} may be reconsidered. In such a case, *the temperature of the corona may not be as high as one million degrees.* In the ionosphere, if we estimate the temperature in terms of the excitation potential of oxygen atoms (4 eV), it would be 45000 K, but it is only about 1200 K (Walker and Rees, 1968). Thus, my question is if the corona is as hot as generally considered.

This is a joint work with Lou Lee in Taiwan.

..

My motive

I have long been interested in the solar corona after I participated the total solar eclipse observation in the South Pacific (Section 2.1). Since then, I leaned about the ionization of the upper atmosphere by energetic current-carrying electrons, which causes the aurora.

When I learned that the high temperature of the corona is not yet solved, I simply thought that what I learned in the auroral ionization by current-carrying electrons may be applicable to the coronal ionization.

Further, I learned that the corona has a large number of loop structures, along which current-carrying electron can flow. From the book by Aschwanden (2005), I found that there is no paper which consider the coronal ionization by current-carrying electrons.

Everyone knows that the corona is somehow related to the aurora (as the source of the solar wind), but no one have thought about another relationship that the ionization process may be similar. Surprisingly, the number density is not far from the upper atmospheric one, so that it seems that the ionization along the coronal loops may partially be caused by current-carrying electrons energized by the double layer.

References

Aschwanden, M., 2005, *Physics of the Solar Corona*, Springer, in association with Praxis Pub., Chichester, UK.

Akasofu, S.-I. and Lee, L.-C., 2019, On the explosive nature of auroral substorms and solar flares: The electric current approach, J. Atmos. And Space Phys., **186**, 104. https://doi.org/10.1016/jastp.02.007. 2011GM001170

Rees, M., 1989, *Physics and Chemistry of the Upper Atmosphere*, Cambridge University Press.

Walker, J. C. G. and Rees, M. H., 1968, Ionospheric densities and temperatures, Planet. Space Sci., **16**, 459-475.

7.3 Solar wind

(a) Generation by the (J x B) force

In this section, we show that electric current might play an essential role in generating the solar wind over the entire heliosphere. The (J x B) force is far greater than the solar gravitational force.

Although the first theoretical paper was published in 1958 by Gene Parker, its cause is still very controversial at best even after more than 60 years (cf. Viall and Borovsky, 2020), in spite of a great number of observations and theoretical studies on it; see Section 1.4 (e) for the history

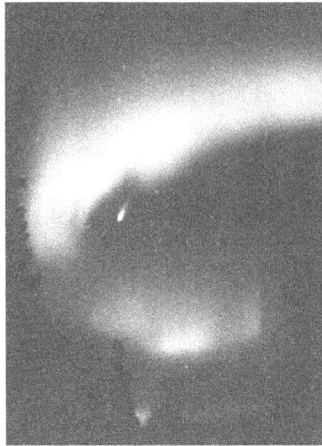

Figure 7.28 Photograph of a comet and the aurora togethe (GI). Both the tail of comets and the aurora are caused by the solar wind.

First of all, the solar wind blows all the way to the edge of the heliosphere (about 100 au), so that the driving force must be present in the entire heliosphere, not just from the bottom in order to overcome the powerful solar gravity (which can hold Jupiter).

Secondly, it is crucial to recognize the presence of a powerful electromotive

force called the Lorentz force (*J* x *B*), not a thermodynamic force alone.

Thus, the problem is to look for an *electric current system, which could* supply such current in the heliosphere and its power supply, the *electric current approach.* Under such a consideration, Lee and Akasofu (2021) considered a large-scale and powerful electromotive force, which is present everywhere in the heliosphere.

For this work, we knew that Alfven (1950, 1977, 1981) suggested that there is *the unipolar induction current system* around the sun. The current system is generated by a rotating dipolar magnetic body like the sun in the heliospheric plasma.

In his model, the electric current flows out from the northern pole of the sun along the polar axis. After reaching the pole of the heliosphere, the current flows along the assumed spherical outer surface of the heliosphere to the equatorial plane and then flows back radially to the sun along the magnetic equator. This current system can reproduce also Parker's spiral magnetic configuration on the equatorial plane as well as in higher latitudes.

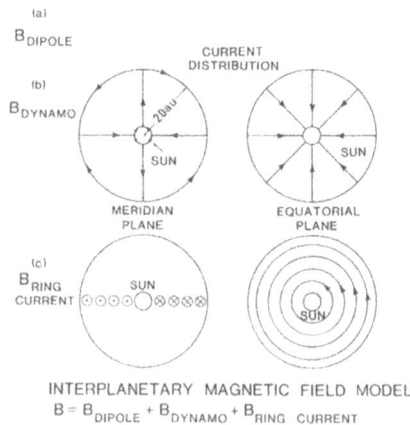

Figure 7.29 Heliospheric current system driven by the solar unipolar induction system.

Alfven (1977, 1981) mentioned that the dynamo process occurs on the photosphere and estimated the dynamo-induced current to be 1.5×10^9 A in one hemisphere on the basis of the observed spiral IMF configuration. The magnetic configuration of the heliosphere can be computed on the basis of the solar unipolar induction current system.

Based on the above model, Akasofu et al. (1980) computed earlier the magnetic field line configuration produced by the unipolar current system in the northern heliosphere. In the past, the heliospheric magnetic configuration was shown only on the equatorial plane or meridional cross sections. Here, we can visualize a 3-D configuration in interstellar space.

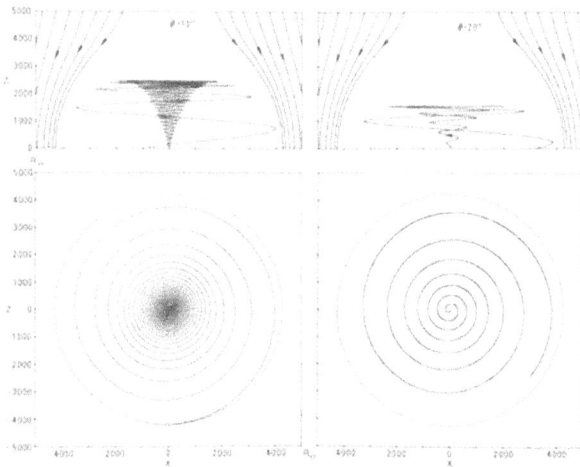

Figure 7.30 Magnetic field configuration of a spherical heliosphere; the field line originated at 10° and 20° from the pole are shown; in this model, they are not connected to the interstellar magnetic field (Akasofu and Covey, 1981).

Akasofu et al. (1980) computed also the magnetic field line configuration on the outer surface of the heliosphere. This has the right configuration, in which the (J x B) force points outward. However, after several attempts, Lee and Akasofu (2021) found that one third of the surface current may be located much closer to the sun (not uniformly distributed in the heliosphere), say at about 10 solar radii.

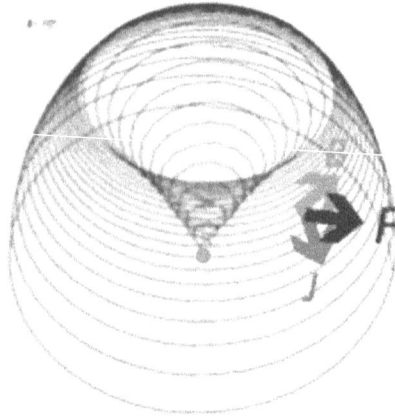

Figure 7.31 A magnetic field line on the outer boundary of the heliosphere, generated mainly by the longitudinal part of the current along the outer surface of the heliosphere. It shows the (J x B) force, which points outward.

They showed that the solar wind speed thus driven at 1 au is 200 km/s under the solar gravitation force. Although we could not reach the speed of 800 km/s, it is hoped that future models of the solar wind will be developed along the line mentioned in the above, because the needed large-scale force is the electromotive force (J x B), which is far greater (*more than 50 times*) than the solar gravitational force at a distance of 10 solar radii in our model.

(b) Latitudinal distribution of the speed

In a study of the solar wind, there is an important observation of the solar wind speed near the sun by the Ulysses.

Fortunately, this observation was made when the sun was extremely quiet (when the ecliptic equator coincides with the magnetic equator). Its *simplicity* in Figure 7.32 suggests that it must be indicating the nature of the basic driving process without being disturbed by sunspot activities. The uniform part of the flow is called *the UF flow* (uniform flow) here.

As can be seen in the following, *we identify that the UF flow is likely to be the flow generated by the basic wind-driving process.*

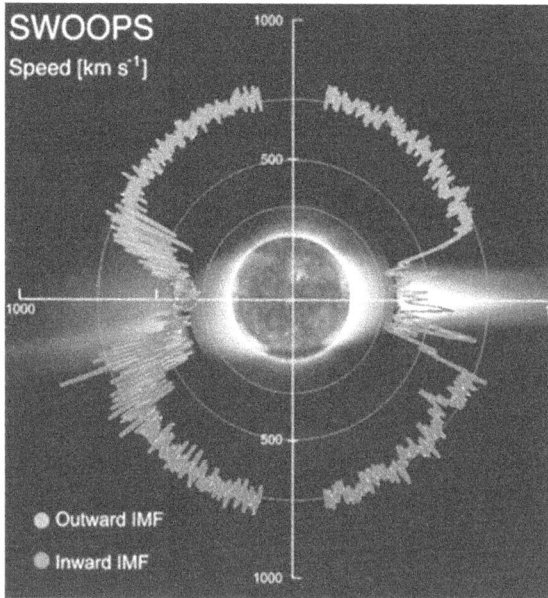

Figure 7.32 The Ulysses observation of the solar wind speed as a function of latitude during the lowest solar activity period. In low latitudes, the wind distribution shows a 'gap', where the equatorial streamer is present (McComas et al., 2013).

(c) Longitudinal distribution of the wind speed

The wind distribution in Figure 7.32 can be shown in a graphic form on the source surface (a spherical surface of three solar radii), called the Carrington map (latitude-longitude, $360° = 27$ days [the solar rotation period seen from the earth]); in this quiet sun condition, the solar magnetic equator coincides with the ecliptic equator. The earth's location projected along a radial line (with the seasonal change of $\pm 7°$) is also shown. The earth scans it from left to right in 27 days as the sun rotates in 27 days. In this case, the earth is confined within the speed region of 400 km/s, so that we can observe the speed of only 400 km/s or less.

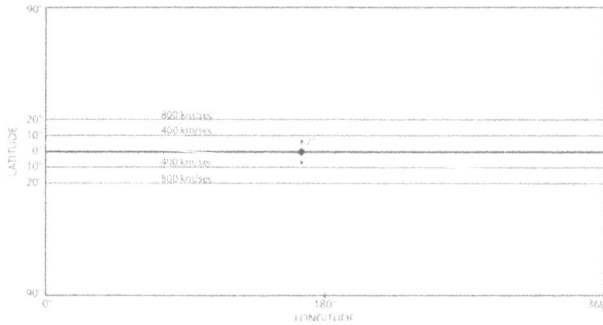

Figure 7.33 Graphic representation of Figure 7.32 observation in the Carrington map (without the streamer). The earth's location is projected with its seasonal change of ± 7°. The seasonal change of the earth's location with respect the ecliptic equator provides higher speed flows and higher geomagnetic disturbances during equinox months (Akasofu and Lee, 2023).

However, the solar wind speed varies considerably from 350 km/s to 800 km/s at the earth's location, suggesting that sunspot activity tends to modify and reduce the speed distribution. It is known that the speed tends to be low during the period of large number of sunspots. We try to explain this contradictory fact in the following.

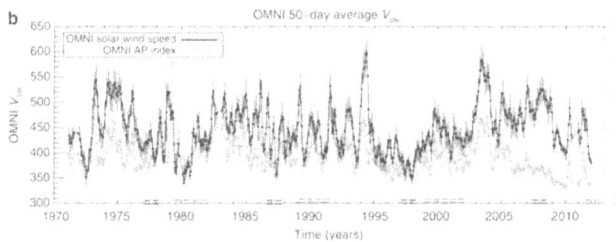

Figure 7.33 OMNI 50-day average solar wind speed between 1971 and 2012. The sunspot maximum period is shown in red, the minimum period in blue (NASA solar wind collection).

In order to understand why the earth encounters such a great variety of the solar wind speed, it is necessary of learn about the magnetic equator of the sun. Saito et al. (1989) projected the magnetic equator on the *source surface* during the whole sunspot cycle 21, which was determined by the records from the Wilcox Solar Observatory (WSO). The magnetic equator can be determined by examining the distribution of sunspots on the Carrington map on the photosphere. The magnetic equator deviates greatly from the ecliptic equator during most of the cycle period. The equivalent dipole axis (based on the magnetic equator) rotates by 180° during the cycle; it is known that the magnetic polarity changes every 11 years, the solar cycle.

Figure 7.34 Solar cycle variation of the magnetic equator on the source surface (the solar cycle 21), based on the neutral line on the photosphere (WSO), together with the sunspot number (Saito et al., 1989).

In spite of its complexity, it can be shown that the magnetic equator can often be represented approximately by a sinusoidal-like curve on the source surface (and the corresponding warped magnetic plane at distances from the sun), particularly after the peak of sunspot cycle. This can be demonstrated by comparing the magnetic equator of CR 1720 during a high solar activity period (1982, sunspot number about 250) with a pure sine wave case. The fact that the sinusoidal magnetic equator can also be understood in terms of the inclination of solar dipole axis; see Figure 7.35.

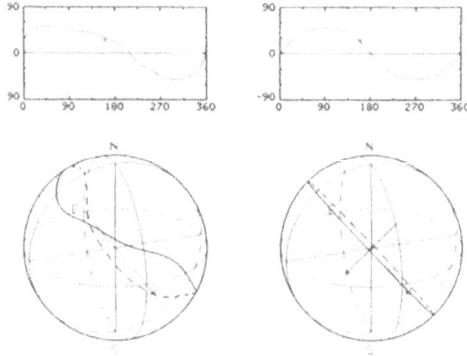

Figure 7.35 Left: An example of the magnetic equator on the Carrington map of CR 1720, both on the rectangular and spherical source surfaces (rotated by 180° for comparison). Right: The simplest sine wave case.

Thus, the magnetic equator can approximately be represented by a sinusoidal-like curve in the Carrington map with varying amplitude (larger for higher activity) when the sun is active. Sinusoidal condition was extensively examined by Akasofu et al. (2005).

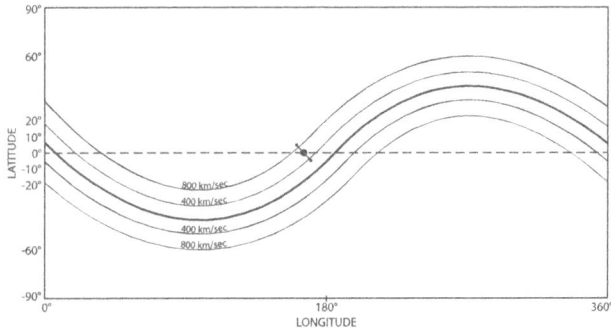

Figure 7.36 The solar wind speed distribution during an active period by making the magnetic equator sinusoidal. The seasonal change of the earth's location with respect the ecliptic equator provides higher speed flows during equinox months (Akasofu and Lee, 2023).

Based on the magnetic equator on the Carrington map during active period of the sun, the earth observes *two long periods of the UF flow of 800 km/s* (one flow is from the northern hemisphere and the other from the southern

244

hemisphere (this is checked by the IMF azimuth polarity) and *two very slow speed between them (350 km/s) for a relatively short period.* Thus, we can understand why the solar wind speed can observed at the earth's location varies during 27 days.

(d) Solar wind when there is no major solar activity (sunspots)

The sinusoidal distribution suggests that we can use the method developed in in Section 7.1 (i).

The simulated solar wind parameters at 1 au by our simulation method mentioned in (i) in this section Figure 7.18b) is compared with the observed one. The simulation can reproduce reasonably well the observed one. In the past, those high speed (750 km/s at the peak) wind have been thought to be the "*streams*" from coronal holes. This simulation shows that *the high speed "stream" is actually the UF flow, namely the flow generated by the solar wind driving process;* Figure 7.37.

Thus, the UF flow generates two co-rotating structures, which cause two recurrent geomagnetic storms (27-day recurrent) in the figure below.

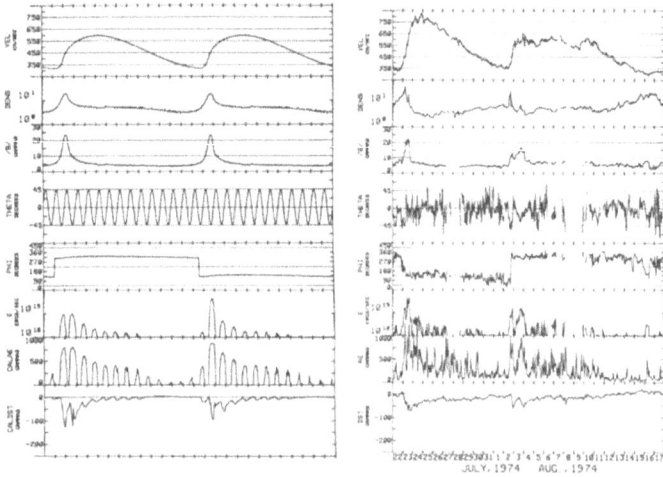

Figure 7.37 Comparison between the simulated solar wind and the observed one. From the top, the solar wind speed, density, IMF magnitude B, IMF angles THETA and PHI (note the change, indicating one from the northern hemisphere and the other from the southern hemisphere). In the simulation, the solar wind-magnetospheric dynamo power ε (= $P/8\pi$), and geomagnetic indices AE and Dst are shown. In the simulation; the THETA angle is represented by sinusoidal variations. The two recurrent storms are reasonably well reproduced (Hakamada and Akasofu, 1982).

It has long been wondered why *polar c*oronal holes can produce a very high speed (750-800 km/s), although there is no particular activity on the photosphere, even if they are open regions; Cranmer (2009). Based on Figure 7.40, we conclude that *the high speed flow may be identified as the UF flow resulting from the sinusoidal-like curve, rather than from coronal holes.* Thus, as mentioned earlier, the highest speed flow is the flow generated by the solar wind generation process.

Thus, based on the above studies, the *"stream" can now be identified as the UF flow, namely the basic flow generated by the solar wind generation process, which produces the 27-day recurrent geomagnetic storms.*

In Figure 7.40, both Dst and AE indices are empirically computed based on the solar wind parameters and are compared with the observed one. The THETA angle is represented by sinusoidal curves. One can see two moderate

246

geomagnetic storms. *Comparing both, it can be seen that the two geomagnetic storms can be reasonably reproduced by the two co-rotating structures.*

From the comparison of the angle THETA (sinusoidal changes), the reproduced AE and the observed AE, one can see that the IMF field lines are continuously vibrating during this period. The amplitude of such a disturbance seems to depend on the magnitude IMF of the co-rotating structure.

Since we can reproduce 27-day recurrent geomagnetic storms in terms of the intensity as a function of time, if the magnetic equator on the source surface is available.

Therefore, we have succeeded in developing the prediction scheme for predicting 27-day recurrent storms on the basis of the solar condition.

Actually, the above results can be seen also in the C9 geomagnetic index which includes the period in the above, together with the recurrent tendency of two geomagnetic storms in one solar rotation period. Since the C9 index is available from the early 1900, we can study changes of the solar wind without interruption.

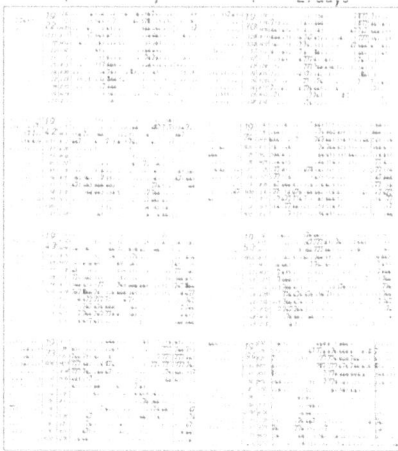

J. Bartels

Figure 7.38 Left: The C9 index during several sunspot cycles; the sunspot situation is shown in the left side of each period. The C9 index is designed to show the two co-rotating structure well. Right: Julius Bartels, who devised the C9 index (as well as the Kp index) is also shown (J. Bartels).

(e) Reproduction of the Ulysses observations during active periods

Here, we demonstrate why the solar wind speed tends to be low when the sunspot number is large.

It is possible to reproduce the fairly quiet and active periods of the Ulysses observations by inclining the most quiet image (Figure 7.32) by adding the equatorial streamer; the image is rotated once around the axis of the sun by considering the observation period. Figure 7.39 shows the comparison during a moderate solar activity, and Figure 7.40 is a very active case. One can see that the basic structure is reasonably well constructed (Akasofu and Lee, 2023).

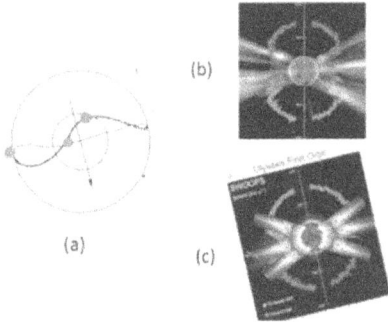

Figure 7.39 (a)An example of the magnetic equator on the source surface during a slightly disturbed condition (near the end of the cycle in this case); The red dots show the assumed location of equatorial streams. (b) The observed distribution. (c) The constructed distribution (b) by inclining Figure 7.32 by about 20°.

In Figure 7.39, (c) is constructed by inclining Figure 7.32 about 20°, putting the equatorial streamer at the location shown in (a) and rotating the sun once. In 7.40, the inclining angle is about 50°.

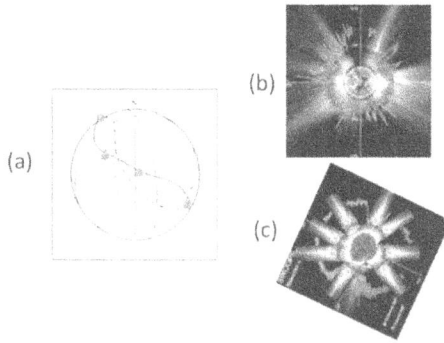

Figure 7.40 (a) An example of the magnetic equator on the source surface (CR 1720, rotated by 180°); red dots show the assumed location of the equatorial streamers. (b) The observed distribution. (c) The constructed distribution of (b) by inclining Figure 7.32 by about 50° and rotating the sun once, putting the equatorial streamer (red dots).

From the above study, one can see why the solar wind speed tends to be low during the sunspot maximum period. The solar wind speed distribution in Figure 7.32 is disturbed by the sinusoidal magnetic equator and the equatorial gap and equatorial streamers (which may not blow to a great distance). On the other hand, we observe the highest solar wind during declining period of the sunspot cycle, because a well-defined sinusoidal distribution of the magnetic equator on the surface and lower sunspot numbers during that period, so that we can observe UF flow well (generated by the solar wind process).

My motive

Although I learned by Alfven's book (1950, 1981) that the (*J* x *B*) force is very powerful, I had no idea how this idea can be applicable to the generation of the solar wind.

When I found his description of unipolar induction in his book (Alfven, 1981), we studied a 3-D configuration of the heliospheric magnetic configuration. However, I did not know how the unipolar induction is related for the solar wind.

Then, on one day some years later, I suddenly recalled that we computed earlier the magnetic field configuration on the outermost surface of the heliosphere; I found that it has the right configuration, which could produce an outward-oriented (J x B) force.

Thus, I contacted immediately Lou Lee in Taiwan (who was a co-author of studying the magnetic configuration of the heliosphere) and suggested to work together to see if the (J x B) force in the heliosphere might generate the solar wind. We could reach the speed of 200 km/s at 1 au as reported in Journal Geophysical Research. It is hoped that someone would continue this work to find it could reach 800 km/s, because the (J x B) force is much greater than the solar gravitational force.

I found the Ulysses observation (Section [b] in this section) of the latitudinal speed distribution (750-800 km/s) and was struck by its *simplicity*. I thought immediately that it must indicate the basic distribution of the solar wind. At that time, the polar coronal hole was believed to be the cause of the highest speed "stream", which caused recurrent geomagnetic storms. Since I do not think that the coronal hole can produce the highest speed stream (because it is the quietest region on the sun, although it might be open), I thought that the UF flow is actually the basic solar wind flow. In fact, our simulation study can reproduce this idea ([d] in this section).

References

Akasofu and Covey, 1981, Magnetic configuration of the heliosphere in inter stelar space, Planetary and Space Sciene, **29**, 313.

Akasofu, S.-I., Gray, P. C. and Lee, L.-C., 1980, A model of the heliospheric magnetic config, Planetary and Space Science, **28**, 609.

Akasofu, S.-I., Watanabe, H. and Saito, T., 2005, A new morphology of solar activity and recurrent geomagnetic disturbances: The late-declining phase of the sunspot cycle, Space Sci. Rev.,120: 27, Doi: 10. 1007/s1124-005-8052-3.

Akasofu, S. I. and Lee. L.-C., 2023, The basic solar wind speed distribution and its sunspot cycle variations, Frontiers in Astronomy and Space Sciences, Doi: 10.3389/fsas.2023.1129596

Alfven, H, .1950, *Cosmical Electrodynamics*, Oxford Univ.Pss

Alfven, H., 1977, Electric currents in cosmic plasmas, Rev. Geophys. Space Phys.15, 272-284.

Alfven, H., 1981, *Cosmic Plasma*, D. Reidel Pub. Co., Dordecht, Holland.

Lee, L. C. and Akasofu, S.-I., 2021, On the causes of the solar wind: Part 1. Unipolar solar induction currents, J. Geophys. Space Phys., **126**, 1, e2021JA029358. https://doi.org/10.1029/2021JA029358.

McComas, D. J., Angold, N., Eliott, H.A., Livadiotis, G., N.A. Schwadron, N. A., Skoug, R. M. and Smith, C. W. (2013). Weakest solar wind of the space age and the current "mini" solar maximum, ApJ **779:2** (10pp).

Viall, N. and Borovsky, J. E., 2020, Nine outstanding questions of solar wind questions, J. Geophys. Res. Space Physics, **125**, e2018JA026005.

7.4 Sunspots

The origin of sunspots is also a long-standing problem. Although many ideas have been proposed in the past, most of them consider processes *under* the photosphere, so that there is practically no way to prove them.

In this chapter, we consider only processes which have been observed on the photosphere. The formation of pore may be most crucial, because single spots are coalescence of pores, which is basically a short, *vertically oriented solenoid of current.*

Historically, sunspots were considered as a crow (three legs) living on the sun by ancient people in China. Galileo was the first to study scientifically sunspots in the early 1700s. William Herschel (1795), the discoverer of the planet Uranus, stated: "(A sunspot) is most probably also inhabited, like the rest of the planets, by beings whose organs are adapted to the peculiar circumstances of that vast globe."

G. Abetti (1955) was perhaps the first person who wrote a comprehensive book on solar physics (Figure 7.41). Chapman gave me his copy.

Figure 7.41 Abetti's book on the Sun (translation).

G. H. Hale (1919; see Abetti, 1955) was the first person who found a strong magnetic field of sunspots and discovered many aspects of the sunspots we know today (Figure 7.42).

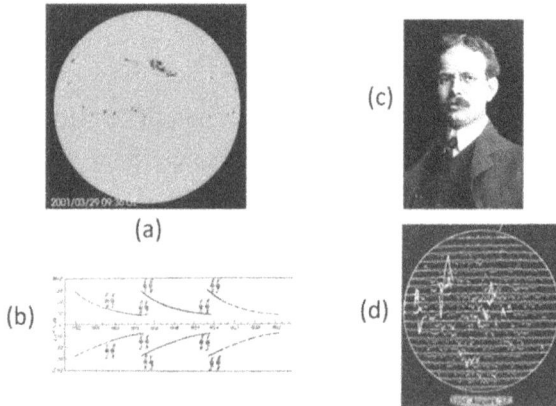

Figure 7.42 Left: (a)An example of large sunspot group. (b) A study by G. H. Hale; it shows one of his important results. (c) G. H. Hale. (d) One of the magnetic scanning of the solar disk by Babcock (all in Abetti's book).

H. W. Babcock was the first person who succeeded in scanning the solar disk by a magnetometer.

We have been taught that sunspots consist of a pair (N&S or positive & negative) like a magnet in our elementary school.

However, I noticed that there are many single spots by scanning many sunspot images at the Kitt Peak Solar Observatory. I confirmed the presence of single spots by reading Abetti's book, in which he shows several such examples. This was the beginning of my inquiry into sunspots as the starting point of my resynthesis of observed facts (not theories) on sunspots (Akasofu, 2020); it is interesting to note that it is only Abetti's book, which describes specifically single spots. As far as I am aware, later books (after the publication of Babcock's paper) do not describe explicitly single spots.

(a) Babcock's theory

There is almost universally acceptance of the theory of the formation of sunspots by Babcock (1961). It is considered that the nonuniform rotation of the sun winds the solar dipole field lines around the sun, forming a thin tube of magnetic flux; when the magnetic tube is supposed to rise above the photospheric surface by magnetic buoyancy, its two cross-sections are thought to be as *a pair of spots*. However, there is no simple way to confirm the theory by observations at the present time. Some solar physicists say that it is a "fairly story", but cannot provide any other theory, when I asked them their alternative.

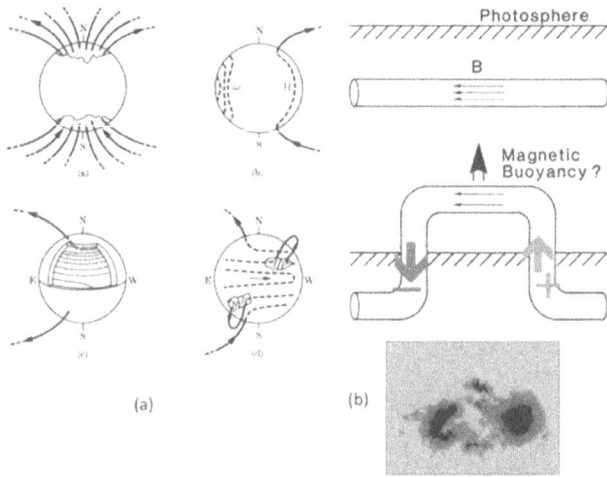

Figure 7.43 (a) It illustrates how the solar dipole field lines might wind under the photosphere by the non-uniform solar rotation. (b) Upper: Schematic illustration of how a magnetic flux tube might emerge from the photosphere, forming a pair of spots. Lower: A typical example of a pair of spots (The Kitt Peak Solar Observatory).

(b) Single spots

There exist a number of spots, which have been called *isolated, independent or solitary spots*. Abetti (1955) named them "unipolar spots" (his plates 99, 100 and 101). In this section, they are called *single spots* by contrasting them with pairs (N/S or P/N) of spots. Thus, single spots *do exist* and are the *simplest* sunspots.

I took this single spot as the basis of developing a morphological theory of sunspots (Akasofu, 2021). High resolution images of single spots show that a single spot consists of several *pores*. There is a gradual change from pores to single spots, depending on the number of pores, so that there is no clear definition of single spots. In fact, the pore may even be the simplest spot.

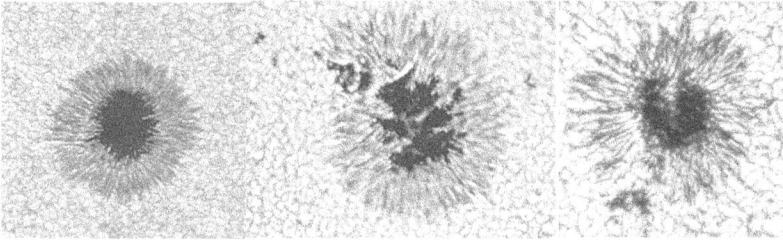

Figure 7.44 Typical single spots (the NASA sunspot collection). The middle and right sides are higher resolution images, showing that a single spot consists of several pores.

It is puzzling why single spots have hardly been studied in the past. This may be because it has so firmly been believed from the earliest days that spots are like a magnet (which has both the N and S poles or P and N together) and because Babcock's theory can so intuitively be understandable.

Another reason may be that "magnetic monopoles" are not supposed to exist in physics, so that single spots are not considered. Thus, single sunspots have almost dismissed or disregarded in the past as a "broken pipe" at best, and I was advised not to consider single sunspots by my solar physics colleagues, who said I was wasting my time.

Incidentally, I was told a story that someone went to Thule, Greenland (near the south magnetic pole), and boiled a large amount of snow, but could not find unipolar (north) magnetic "dust."

Nevertheless, single spots, the simplest form of sunspots, *do exist*, so that it is best to start with a synthesis (a new morphological study) from the simplest case.

This fact that single spots exist is contradictory to the widely accepted Babcock's theory, because sunspots are considered to appear as a pair of positive and negative spots. Thus, the problem I faced was:

When a well-accepted theory was found to be contradicting with observed facts (the existence of single spots) what could I do?

(c) Synthesis

Since I have no experience of observing sunspots in detail myself, this study is based on published papers, books and available data. There are a great number of observed data, so that my task is synthesizing the data.

Unipolar magnetic region

In order to investigate single spots, it became necessary to reexamine magnetic fields on the solar disk as a first step of synthesis (Akasofu, 2014, 2015).

First of all, when one examines the photospheric magnetic field distribution on the solar disk, one can recognize weak positive and negative bands, aligned alternately in longitude (including the northern and southern Polar Unipolar Regions). They are known as *unipolar magnetic regions*.

Figure 7.46 The magnetic field distribution on the photosphere (The Kitt Peak Solar Observatory).

It is generally considered that unipolar regions are old active regions stretched out by the nonuniform rotation of the sun (Leighton,1969). This idea has been well accepted today; Figure 7.47.

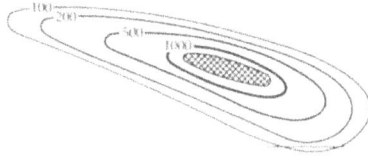

Figure 7.47 It was thought that unipolar regions were old active sunspot region stretched out by the non-uniform rotation of the sun (R. G. Giovanelli).

However, I found, first of all, that unipolar regions grow from their initial location at about 30°, and that their poleward end extends to about at 60° (which is far above sunspot regions); in fact, they reach the Polar Unipolar Regions at a certain epoch of the sunspot cycle and change the polarity of the solar dipole field (Hathaway, 2010); Figure 7.48. They extend also to the equatorward extent (0°) as the sunspot cycle develops.

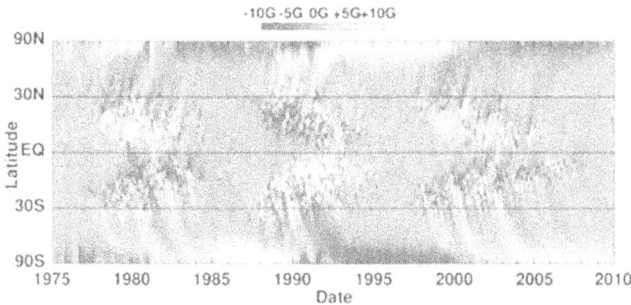

Figure 7.48 The extension of the polar end of unipolar regions to the polar unipolar region, which changes the polarity of the solar dipole field (Hathaway, 2010).

Further, *they grow before sunspots appear at the beginning of the cycle, and sunspots grow and decay within unipolar regions*. Thus, it is likely that they are closely related to the intrinsic internal dynamo process, which causes the 11-year solar cycle. These facts disagree with the existing idea of unipolar regions (Leighton, 1969). I am not aware any specific theory of unipolar regions other than by Nakagawa (1971).

Thus, I had to face the two major contradictions at once, disagreement with Babcock's theory and also Leighton's theory of unipolar regions.

When we examine further the magnetic distribution on the solar disk, we can see many concentrated fields in unipolar regions. They are pores and single spots.

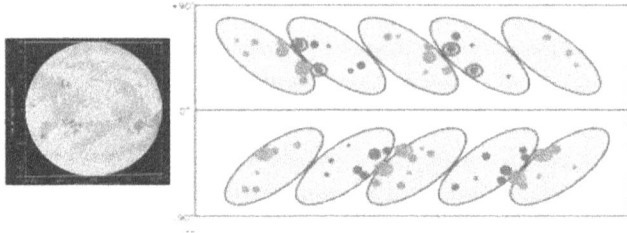

Figure 7.49 Left. The magnetic distribution on the solar disk. Right: Schematic illustration of the left in the Carrington map (longitude-latitude); a small dot (with various sizes) indicates a single spot. Large dots across the boundary of unipolar regions are a pair of spots.

The most important point here is that positive concentrated field (bores and single spots) are located on positive unipolar regions (vice versa); (Akasofu, 2014, 2015). Figures7.49 and 7.50 show schematically this observed fact.

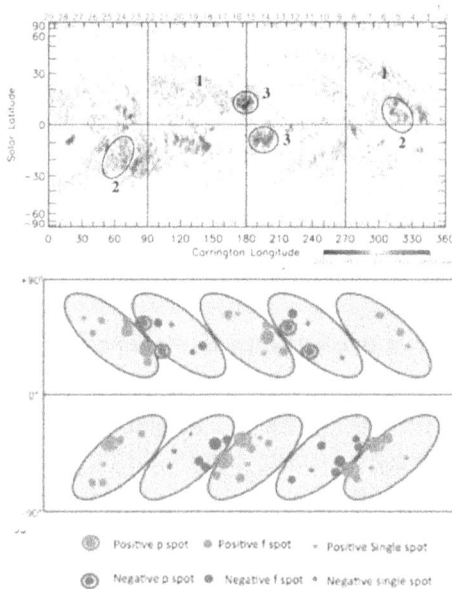

Figure 7.50 Upper: An example of the observed magnetic field distribution on the

Carrington map in both hemispheres and its schematic view. Unipolar regions are numbered as "1". Single (unipolar) spots are denoted by "2". Pairs (positive and negative) are designated as "3"; they are located at the boundary of positive/negative unipolar regions (Kitt Peak Solar Observatory). Lower: A schematic illustration of the upper figure.

As mentioned earlier, unipolar regions *grow before sunspots appear at the beginning of the cycle, and sunspots grow and decay within unipolar regions*; Figure 7.51. Thus, like the solar dipole field, it is likely that they are related to the intrinsic internal dynamo process. These facts disagree again with the existing idea of unipolar regions.

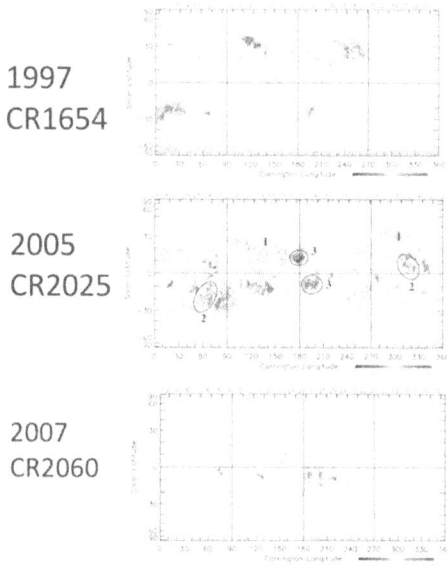

1997
CR1654

2005
CR2025

2007
CR2060

Figure 7.51 Sunspot Cycle variation of the unipolar regions. They grow and decay with the sunspotcycle, so that they are not old active sunspot groups.

(d) Pores and single spots

The photosphere is covered by convection cells, so that irregularities (non-uniformity) of the convective motions leave some magnetic fields at the boundary of the convection cells, forming *pores*. They may be associated with small eddy or vortex flow (Parker, 1992).

The figure below shows schematic view of negative pores and single spots, which are formed within a negative (blue) unipolar region.

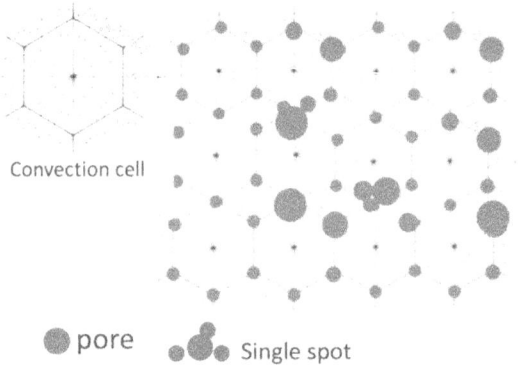

Convection cell

● pore   ●🟤● Single spot

Figure 7.52 One convection cell considered by Clark and Johnson (1967) is shown in the upper left. A schematic illustration of a negative unipolar region by small dots (high space resolution) with pores and single spots.

Positive pores are formed in a positive unipolar region (vice versa), and a positive single spot is formed by coalescence of positive pores.

Norlund et al. (2000) made an extensive study of granulation; they made a detailed study of the relationship between the convection and magnetic fields. Recently, Bharti et al. (2016) examined in fine details of the relationship between granules and pores, umbral and penumbral structures, including flows toward pores.

Their result seems to show that pore is a short column of magnetic flux of vertical structure of length of a few Mm; its diameter is less than 10^3 km. Thus, one can consider a short and vertical solenoid of electric current.

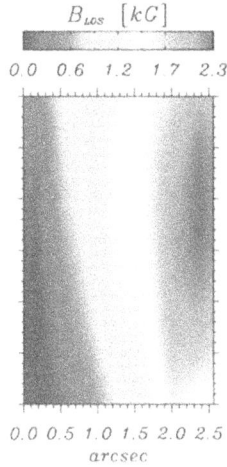

Figure 7.53 Magnetic field of pore in the vertical cross section (Bharti et al., 2016).

(e) Monopole problem

As stated earlier, the important point is that a positive single spot is always formed in a positive unipolar region (vice versa).

Since this local convergence/coalescence process occurs only in small and lowest areas within unipolar region, it is likely that there is no significant magnetic flux re-arrangement in its upper part of unipolar magnetic regions or in neighboring unipolar magnetic regions. This situation may be similar to that envisaged by Cranmer (2009) (although he produced it in a different context); Figure 7.54.

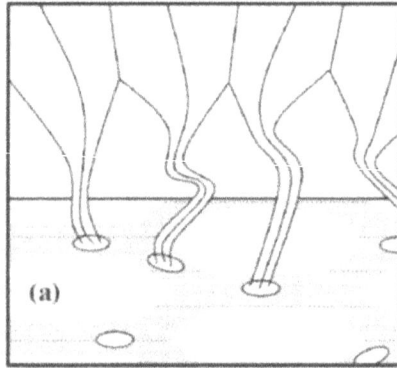

Figure 7.54 The local convergence of photospheric magnetic flux, taken from Cranmer (2009, his figure 4), in which he used it for a different context.

This consideration of the formation of pores and single spots may solve the problem of a single spot as a "magnetic monopole," and thus there is no need to look for its counterpart in a variety of forms in other areas. The point here is that a positive single spot is formed by local convergence in a positive unipolar region.

High resolution images of single spots show that a single spot is coalescence of pores.

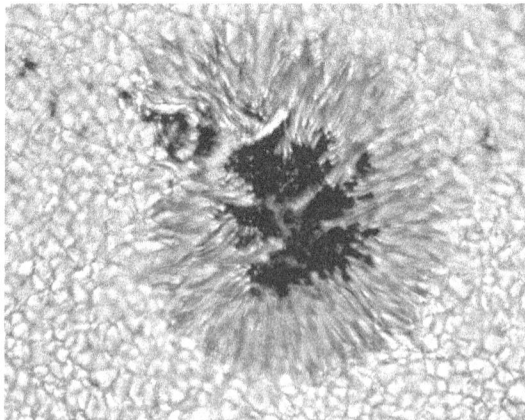

Figure 7.55 An example of single spots, which are an assembly of pores (NASA Sunspot Collection).

(f) Plasma flows around a single spot

(1) Outward from the top

This flow is the well-known Evershed flow out from the top.

(2) Downward flow from the top along the side of a spot (Sodanki. 2003).

(3) Converging and diverging flows near the bottom (some contradicting reports). There are many subsurface observations of flows under active regions; they are results of helioseismology (cf. Haber et al., 2004; Hindman et al. 2009; Lagg et al. (2014); Komm and Gosain, 2015).

(g) Electric currents around a single spot

If a pore is like a short column of magnetic field, a single spot (as a coalescence of pores) must also be surrounded by *electric current.* It is possible to consider a vertical column of circular current or a short solenoid of length of a few Mm (not the long flux tube surrounding the sun, which was considered by Babcock).

Indeed, Kotov (1971) showed a circular currents of 10^{12} A around a single spot; Figure 7.56. Solanki (2003) mentioned that the current intensity is 6-7 mA/m^2..

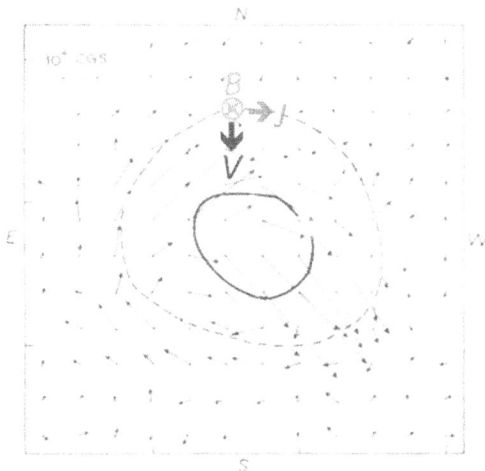

This current must be associated with the plasma flow pattern (*V*) in Figure 7.56. The origin of the inflow is discussed below.

The inward flow generates a clockwise current (*V* x *B*). This inward flow does not depend on the polarity of spots. A positive pore in a positive unipolar region produces also an inward flow. The upward Evershed flow flows out from the top of spots, which does not depend on the polarity of spots.

Assembling all those observed facts, the formation of pores seems to be most basic.

(h) Formation of pores

It seems that pores have a short column of magnetic flux (Bharti et a., 2016). Figure 7.57 provides some hints on the formation of pores. A sunspot is known to have a lower temperature than the surrounding. A slight pressure differential pressure may induce a weak inward flow, when pressure in the pore is not balanced. This inward flow seems to be the first process of the formation of pore and single spot.

In Figure 7.57, the basic flows are shown by red arrows. *Thus, a single spot seems to have a cyclonic plasma flow.*

If this process is too weak, pores may not develop (as often the case).

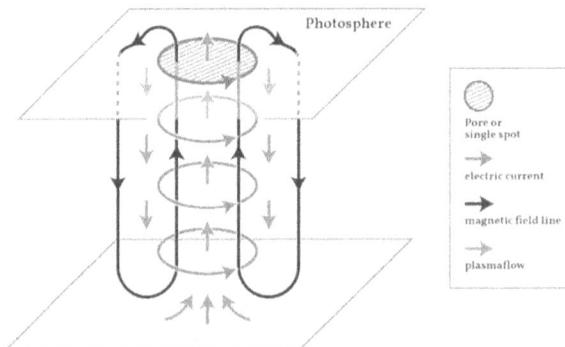

Figure 7.57 A tentative model of a positive pore or single spot which is constructed by synthesizing the observed facts. For a negative pore or single spot, the direction the electric current is reversed, but *the direction of the plasma flow is the same.*

(i) Formation of single spots

Since single spots are produced by coalescence of pores, the question is what force can bring pores together. Parker (1992) has an interesting paper in this regard, titled "Vortex attraction and formation of sunspots". It is best to quote his abstract: "A downdraft vortex ring in a stratified atmosphere exhibits universal attraction for nearby vertical magnetic flux bundles. We speculate that the magnetic fields emerging through the surface of the Sun are individually encircles by one or more subsurface vortex rings, providing an important part of the observed clustering of magnetic fibril to form pores and sunspots."

The dynamics of the coalescence of pore is still a major problem as shown in the following.

(j) Cyclonic structure

McIntosh (1981) suggested that sunspots show a cyclonic feature. This feature has not got enough attention, perhaps because of the fact that the sun's rotational speed is only just comparable with various photospheric motions. Thus, it is likely that there is a cyclonic motion around single spots and large spots. Figure 7.58 shows an example (Min and Chao, 2009).

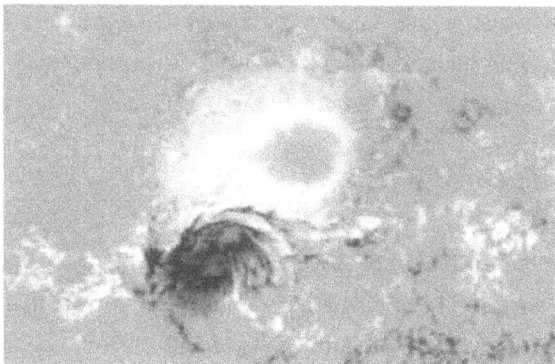

Figure 7.58 An example of intense cyclonic motion of a spot (Min and Chao, 2009).

Thus, an extensive search is made to find the cyclonic feature from a large number of sunspot photographs taken at the Kitt Peak Solar Observatory. Figure 7.59 shows an example, in which the Coriolis force around a sunspot (in the norther hemisphere) is so clearly exhibited (Akasofu, 1985).

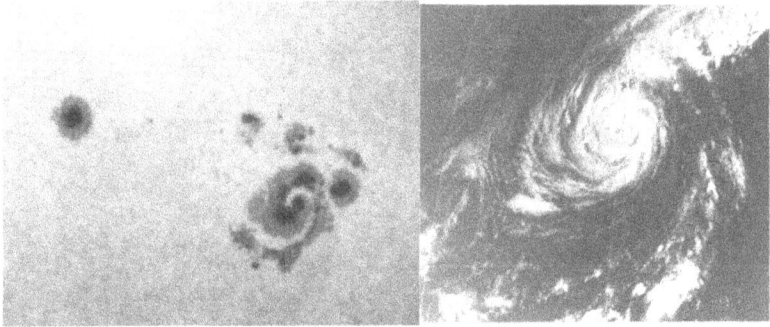

Figure 7.59 Left: Cyclonic structure of sunspots. There are several pores (Akasofu, 1985); (Kitt Peak Solar Observatory). Right: Hurricane (typhoon) near Japan (Japan Meteorological Agency).

Associated with the above cyclone structure of sunspot, Gene Parker and I corresponded on the formation of sunspots. He wrote to me in his letter, dated on 27 June 1984:

"Dear Syun Thanks for the reprint and preprint. We have been thinking along similar lines so far as the emergence and disappearance of Ω-shaped flux tubes [Babcock's theory] is concerned. I enclose a reprint of my paper on the subject, 'Depth of Origin of Solar Active Regions'. I was interested to read the preprint on vertical sunspot structure. The photograph is striking. I have for some time argued that one can understand the formation of sunspots only in terms of a convergence flow and associated downdraft. ---. Gene".

(k) Torsional oscillation

Any theory on the formation of sunspots has to explain the equatorward shift of sunspots during the sunspot cycle. It is interesting to note that the torsional oscillation zone shifts equatorward (Howard and LaBonte, 1980). It may be speculated that the interaction between the unipolar region and the torsional oscillation are also related to the formation of sunspots. In fact, there occur many solar flares along the line of the torsional oscillation (see Figure 7.60).

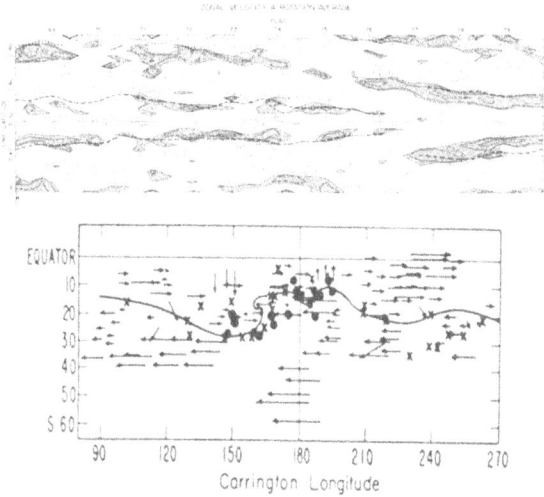

Figure 7.60 Upper: The equatorward shifting of the torsional oscillation (Howard and La Bonte, 1980). Lower: The torsional oscillation belt (arrows) and solar activities (flares) indicated by dots. It shows that various solar activities, including the formation of spots, occur along torsional oscillation (McIntosh, 1981).

(l) Pair of spots

As shown earlier, *a distinct pair of spots forms only at the neighboring unipolar regions (positive and negative)* [Akasofu, 2014, 2015]. A pair of spots is *not* formed in the middle of unipolar regions. Secondly, a "spot" of the pair consists of is a cluster of single spots by coalescence (McIntosh 1983). Obviously, from what we have learned earlier, a positive cluster is formed in positive unipolar regions (and vice versa).

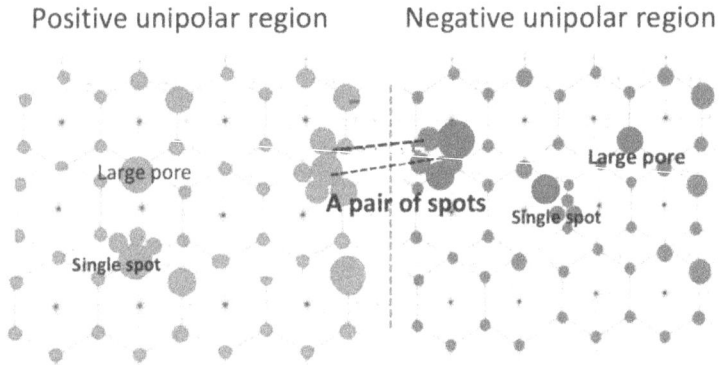

Figure 7.61 A schematic illustration of a pair of spots at the boundary of neighboring (P/ N) unipolar regions, showing one cluster in each side of positive and negative unipolar regions. They are connected by magnetic field lines.

Further, what is generally called 'a pair of spots' is actually a pair of clusters in high-resolution images (Figure 7.62). In some high-resolution images, each side of a spot consists of a large number of pores and single spots. They seem to be almost individually connected by magnetic field lines to the counterparts on the other side of unipolar region.

Figure 7.62 An example of "a pair of sunspots", which is actually a pair of clusters of single spots (NASA sunspot collection).

From these observations, it may be speculated that a positive cluster in a positive unipolar region and a negative cluster located near the boundary of unipolar regions form a pair. For this speculation, there is an interesting observation by Sheeley (1976) on magnetic field lines.

Two clusters across the boundary of neighboring unipolar regions are often connected by magnetic field lines *above* the photospheric surface. It is likely that the magnetic linkage between two active segments occurs in the way Sheeley (1976) presented; on the basis of X-ray images of magnetic field lines, he described: "---*these field lines usually interact by changing their flux linkage, much as they do in a vacuum*". That is to say, for a given magnetic field distribution of two active clusters at the boundary of neighboring unipolar regions in the photosphere, the magnetic field line linkage may occur almost like in a potential field.

Figure 7.63 shows that a few sunspot *pairs* are magnetically connected among themselves; pairs are likely to be connected even across the equator. Thus, the pattern is very different from what we expect from Babcock's theory. Indeed, the observed field lines on the photosphere are supposed to be below the photosphere (invisible) in Babcock's theory.

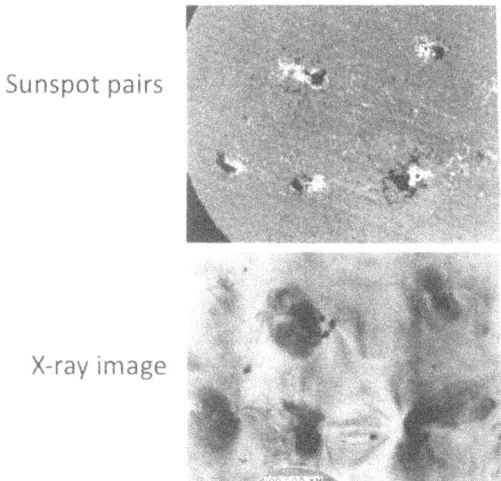

Sunspot pairs

X-ray image

Figure 7.63 Upper: a few pairs of spots in the northern and southern hemispheres across the equator. The X-ray image shows that Babcock's magnetic flux (Figure 7.43) is above the photosphere, not below the photosphere as he suggested (Sheeley,1976).

(m) Summary

It is shown that a study of single spots has led us to four new findings.

(1) Unipolar magnetic regions are related to the intrinsic solar magnetism, which grow and decay with the sunspot cycle, not just decaying old active regions.

(2) Positive single spots are formed in positive unipolar regions by coalescence of positive pores (vice versa).

(3) A pair of spots is formed at the boundary of neighboring (P/N) unipolar regions, not in the middle of a unipolar region.

(4) *A dense cluster of single spots (and large spots) is formed at each side of neighboring (positive and negative) unipolar regions, not in the middle of unipolar regions.*

Based on these findings, an attempt is made to synthesize a number of observations related to the formation of pores, single spots and clusters of single spots, in addition to the formation of a pair of clusters at the boundary of unipolar regions.

Based on these observed facts, the formation of pores may be:

(a) The convective motion of cell associated with the photospheric convection in a positive unipolar region leaves a weakly concentrated positive magnetic field flux (vice versa).

(b) The differential pressure with the surrounding causes converging flow, concentrating a short magnetic flux, forming pore.

(c) There is a cyclonic flow around a spot.

(d) Several pores coalesce to form a positive single spot.

(e) Single spots may coalesce to form a larger positive 'spot'.

(f) When a coalesced single spot appears near the boundary of unipolar regions (positive/negative), both forms (positive and negative) pair.

Based on the above points of view, two fundamental questions are:

(i)   The cause of unipolar regions

(ii) The formation of pores

(iii) The coalescence process of pores/single spots

It is my hope that my synthesis might become a new morphological model or theory.

My motive

When I was scanning films of sunspots at the Kitt Peak Solar observatory, I was surprised by the variety of appearance of sunspots. One of them is single spots. Since we are taught that sunspots consist of a pair (N/S or P/N) of spots, I was surprised by their presence. Further, I was struck by its simplicity compared with many others.

Obviously, single spots are unusual, since we learned that a sunspot consisted of (N/S) pair. I tried to find papers or recent books on single spots, but could not find about them. Fortunately, I found them in the first comprehensive (old) book on the sun by Abetti (1955); he described them as solitary or independent spots.

In facing the well-established paradigm (Babcock's theory) in my sunspot study, my first step was to examine a large number of data set. Then, I found that there are a few observations which do not seem to be consistent with the paradigm theory.

1. *Examination of a large number of sunspots.*

2. *Accidentally finding single spots (Figure 7.48).*

3. *Searching for papers or books on single spot (Figure 7.45).*

4. *Finding that they have not considered in the past.*

5. *Assemble a large number of observed facts on sunspots.*

6. *Finding a few new (or forgotten) fact.*

7. *Construct a new idea of the formation of single spots (Figure 7.60).*

8. *Explaining the formation of coupled (N/S) spots (Figure 7.64)*

9. *Searching for supporting observed facts (Figure 7.66).*

10. *Consulting with my idea with sunspot experts.*

11. *Deciding the publication.*

Episode

(1) Visiting solar observatories

I was interested in visiting solar observatories, such as the Big Bear Observatory in California, the Kitt Peak Observatory in Arizona and the Sunspot Observatory in New Mexico. In the Sunspot Observatory, I had an opportunity to observe solar flares in progress (Hα observation). I found that the development of solar flares is rather slow compared with auroral substorms (less explosive); we had even a little time for a coffee break. Time-lapsed flare films are somewhat misleading.

At the Kitt Solar Observatory, I encountered strong thunder lightnings. At the Big Bear Observatory, I enjoyed discussion with solar physicists and learned why the observatory is set in the middle of the lake (the air is very quiet in the morning). I wanted to visit the historic Mt. Wilson Observatory, but did not have a chance.

(2) Circular flare

There are a great variety of flares. One of them is a circular flare emission. Further, this flare has the *emission on sunspot*, too, which is not rare. One possibility is that a circular photospheric flow constitutes a dynamo process

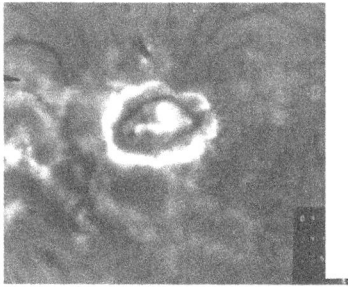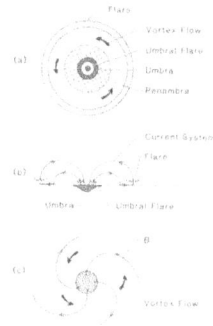

(a)   (b)

Circular flare. Note also the emission on a sunspot.

(3) Visiting Taiwan

I was fortunate enough to have many solar and space physicists in Taiwan.

With Lou-Chaung Lee and Bob McCoy (director of the Geophysical Institute of the University of Alaska Fairbanks) at the Institute of Earth Sciences. Lou and I have had a long association in working on various studies.

References

Abetti, G., 1955, *The Sun*, Faber and Faber, London.

Akasofu, S.-I., 1985, Vortical distribution of sunspos, Planet. Space Sci., **33**, 275.

Akasofu, S.-I., 2014, Single spots, unipolar magnetic regions, and pairs of spots, Geophysical Res. Lett., **41**, Doi:10. 1002/2014GL060319.

Akasofu, S.-I., 2015, Single spots, unipolar regions and pairs of spots, Geophys. Res. Lett., **42**, Doi: 1002/2014GL062887.

Akasofu, S.-I., 2021, A morphological study of unipolar magnetic fields and the relationship with sunspots, J. Atmos. And Solar-Terr. Phys., **218**, 105625, http//doi.org/10.1016/j.jastp.2021.105625.

Babcock, H. W.,1961, The topology of the sun's magnetic field and the 22-year cycle, ApJ., **133**, 572.

Bharti, C. et al., 2016, Quintero Noda, C., Joshi, C. Rakesh, S. and Pandya, A., 2016, Fine structures at pore boundary, Mon. Not. Roy. Astro. Soc., **462**, L93.

Bray, R.J., and Loughheard, R.E.,1964, *Sunspots*, Dover.

Clark, A. and Johnson, A. C., 1967, Magnetic field accumulation in supergranules, Solar Phy., **2**, 433.

Cranmer S. R., 2009, Coronal holes, Living Rev. Solar Phys.,6, 3, Speed during the minimum between Cycles 23 and 24, Solar phys., **274**,

Gizon, L. and Birch, A.C.,2005, Local helioseismology, Living Rev. Solar Phys. **2**, 6, http://www.livingreviews.org/lrsp-2005-6.

Haber, D. H. et al., 2004, Organized subsurface flows near active regions, Solar Phys., **220**, 371.

Hathaway, D. H., 2010, The solar cycle, Living Rev. Solar Phys.,7, 2010, https://www.Livingreviews.org/lrsp-2010-1.

Herschel, W., 1795, On the nature and construction of the sun and stars, Phil. Trans. Roy. Soc. London, **85**, 46.

Hindman, B.W., Haber, D. A. and Toome, J., 2009, Subsurface cicuration s within active regions ApJ, **698**: 1749, doi; 10.1088/0004-637X/698/21749

Howard, R. and La Bonte, B. J.,1980, The sun is observed to be a torsional osccilation with a period of 11 years, ApJ, **239**: L33.

Ji, K. et al., 2016, Investigation of umbral dots with a new vacuum solar telescope, Solar Phys., **291**:357, DOI: 10: 1007/s11207-015-0796.

Khomenko, E. et al., 2015, Evershed flow observed in neutal and singly ionized ion lines, Astronom. Astrophys., DOI: 10. 1051/0004-6361/201526437

Komm, R.K. and Gosain, S., 2015, current and kinetic helicity of long-lived activity complex, ApJ, **798**: 20, (13pp) doi:10. 1088/0004-637X/798/1/20

Kotov, V.A., 1971, 213, IAU Symposium No.43, held at the College de France, August 31-September 4, 1970.ed by Howard, R., D. Reidel Pub. Co., Dordrecht-Holland.

Lagg, A. et al., 2014, Vigorous convection in a sunsot granular light bridge, Astronom. Astrophys., DOI: 10. 1051/0004-6361/201424071.

Leighton, R. B., 1969, A magneto-kinematic model of the solar cycle, ApJ,**156**, 1.

McIntosh, P. S., 1981, The Physics of Sunspots, **Vol.7**, ed by Crom, L.E. and Thomas, J.H., 7, Sacramento Peark Observatory, New Mexico.

Nakagawa, Y., 1971, A numerical study of the solar cycle, Solar Magnetic fields, Howard, R. (ed), IAU Syumosium No.**43**, D. Reidel Pub. Co., Dordrecht-Holland, 725, held at the College de France, August 31-September 4, 1970.

Norlund, A., Stein, R. F. and Asplund, M., 2000, Solar surface convection, http://creativecommons. Org/licenses/by-nc-nd/3.0/de/

Parker, E. N., 1984, Private communication dated on June 27, 1984.

Parker, E. N., 1992, Vortex attraction and formation of sunspots, ApJ., **390**, 290.

Rieutord, M. and Rincon, F., 2010, The sun's supergranulation, Living Rev. Solar Phys.,**7,** 2-82. http://www.livingreviews.org/lres-2020-2.

Saito, T., Oki, T. and Akasofu, S.-I., 1989, The sunspot cycle variations of the neutral line on the source surface, J. Geophys. Res., **94**, 5453.

Sheeley, N. R. Jr., 1976, Energy released by the interaction of coronal magnetic fields, Solar Phys., **47**, 177-185.

Solanki, S. Y., 2003, Sunspots: An overview, Astron Astrophys. Rev.,**121**: 153-7286. Doi: 10. 1007/s00159-003-0018-4.

Svestka, Z., 1958, *Solar Flares*, D. ReiAppendixdel Pub. Co., Dordrecht, Holland.

Chapter 8 How to make an advance in your field: My scientific methodology

—————————————————————

When one begins to study in a particular field, most advanced fields in geosciences, including space physics or solar physics, there is a well-accepted theory or idea. It is often called '*paradigm*'. Since everyone must be interested in making a significant or stepwise contribution in this situation, the question is how to proceed in facing a paradigm.

I gave a talk under the title: "How can you make a ste-wise advance in your field?" at the Geophysical Institute of the University of Alaska Fairbanks in October, 2024. The written version of my talk is presented here as an introductory note of this chapter. It may be useful before reading the main subject.

The title of my talk was "How can you make a stepwise advance in your field ?" I do not have any secret to offer. My hope is to share with you my thoughts on this subject based on my experience during 60 years of research life, which began in 1959.

The main scheme of my talk may be expressed in a simple way in the figure below. When you are trying to solve the Garfield puzzle (which all other people are also working on it), you *happen to encounter* an 'odd' piece that does not seem to belong to Garfield puzzle. When you examine the puzzle further, you find more 'odd' pieces. Thus, you wonder what puzzle you are working on. Eventually, you will find that the puzzle you are working may not be Garfield puzzle, but Snoopy puzzle. Such a process has occurred several times in each field in science. It is such an inquiry mind, which makes a new advance in the field. This theme will be repeated many times in my talk.

Natural phenomena are so complex that there are always 'odd' observational facts, when you are working under the prevailing paradigm (Garfield puzzle). This is the situation every researcher faces, because you are working on the basis of what you have learned.

Creativity

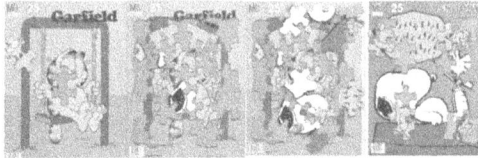

When you study the history of development of Earth sciences, you will find that a new field emerged when a discovery of new phenomenon was made. A new discovery is considered to be a sort of 'odd' piece. The discovery is followed by a variety of hypotheses and models.

On the basis of many observed facts, including an 'odd' piece, one model among many becomes popular. The Garfield model is a model which has been born in this way. If it can survive for a few decades (for several generations), it becomes the foundation of the field, a new paradigm.

Unfortunately, everyone forgets that you are working on such a *model*; it is simply one of the models, and you are trying to improve it. This improvement of paradigm is like arguing if Mona Lisa should have bigger eyes, or a prettier mouth. It is the last clean-up.

All of us spend much of our research time in this process, since we are bound to a specific idea (what you learned). At this stage of the development of field, we believe that the popular model is basically correct, so that what is needed is improving it. We forget that it is simply a model.

Examples of such a model are:

Plate movement by mantle convection; researchers are now studying how earthquakes occur and how volcanos erupt on the basis of plate movement. This knowledge is now commonly accepted and even school children are aware of it.

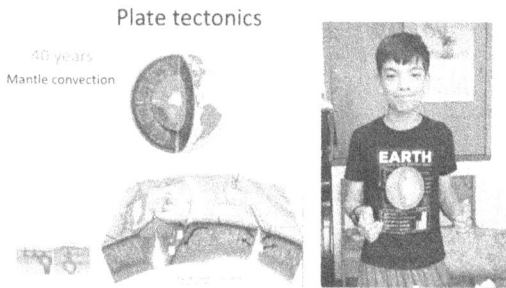

Plate tectonics

Mantle convection

Another example is global warming. Considering that carbon dioxide is its cause, hundreds of simulations are being made to predict the amount of temperature increase by 2100, and many observations of warming (such as glacier retreat and reduction of ice in the Arctic Ocean) are studied and reported almost every day. The great ocean circulation originating from the Arctic Ocean is also such an example.

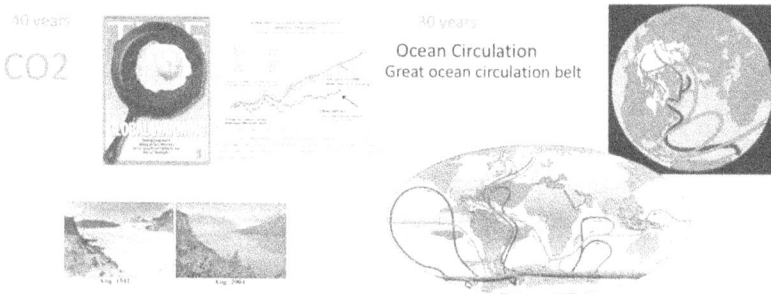

CO2

Ocean Circulation
Great ocean circulation belt

In space physics, from about 1958, researchers have been working on the assumption that an anti-parallel magnetic field configuration in the magnetotail or sunspots explosively annihilates itself and becomes the energy source of auroral substorms and solar flares.

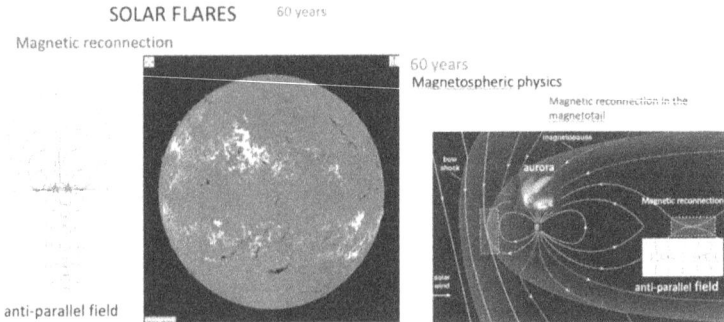

All these models we are currently working on have survived for a few or several decades. Thus, they are taught generation after generation, so that they become the basis for us to work.

If a paradigm lasts a few decades, it is basically accepted without anyone questioning its validity. As a result, it tends to last and becomes even an obstruction for future advances.

In order to explain in more details each of the above examples, let's return to Garfield puzzle.

The first rule in the Garfield puzzle is that the researchers are supposed to consider only the Garfield puzzle, not any others; this situation will become clear later. Second is not to force the piece where it should not be located; for example, considering every anomalous weather event is caused by global warming. The third is that one should not use a pair of scissors (modifying data).

I was told that the word paradigm is a Greek word, which was introduced in science by Thomas Kuhn, a science philosopher. I was also told that it means something like a problem at the end of chapter in a textbook. If you cannot solve it, you feel that you are not capable.

However, it may be that something is not right in the textbook. In fact, if you find that everything in the textbook is correct, what research you have been doing?

If you find your book is basically correct, but it needs some improvements, you have certainly contributed to your field.

However, if you hope to make a stepwise advance in your field (what every scientist hopes), what can you do? This is the subject of my talk.

First step

When you are working on Garfield puzzle (a large amount of observed data), you may *happen to notice* an 'odd' piece that does not seem to belong to the Garfield puzzle (such as an observed fact that does not belong to your paradigm). When you happen to find such an 'odd' piece, there are three choices: (1) Throwing it away, (2) Trying to look for a spot where it might fit, (3) Finding other 'odd' pieces and doubting that you are working on Garfield puzzle.

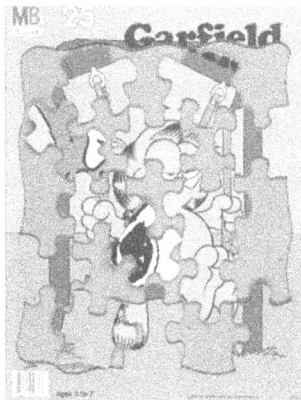

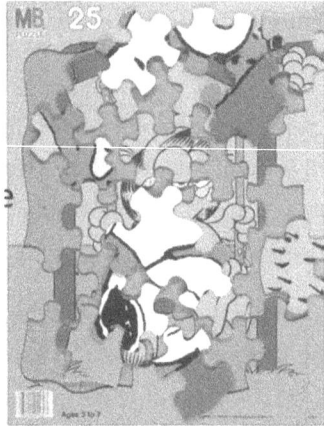

I classify three types of scientists in this regard.

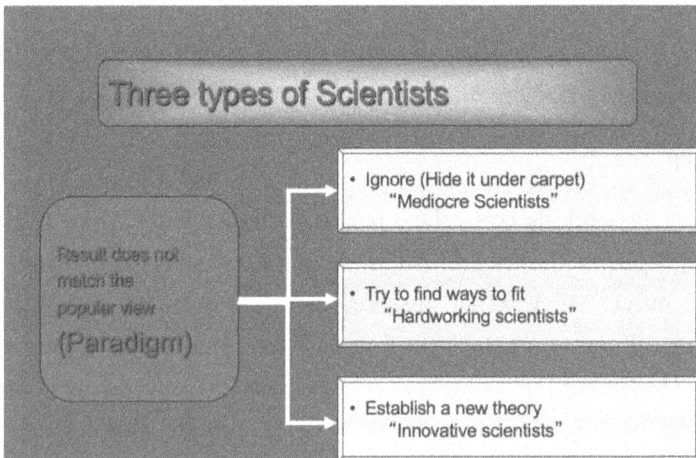

This classification does not depend on the capability of scientists. When you encounter an 'odd' piece, going from the first type to the third type is nothing but the degree of *courage*. In fact, you need much courage to become the third-category scientist.

Second step

Before going from the second to the third step, it is essential to find more 'odd' pieces. If you could find at least several of them, the first odd piece is likely to be *genuine*.

Third step

Now, you have many 'odd' pieces. You have to consider what kind puzzle of animal (or something other than an animal) it is.

You have to synthesize many pieces to come up with the solution.

This process requires *creativity.*

There is a clear definition of creativity in science.

"Creativity in science consists of perceiving a new thought pattern on the basis of already available data or theories." Many people consider that creativity is creating something from nothing and give up creative action.

The point here is *"already available."* Creativity is not the ability to create something from nothing. Many people consider that creativity to produce some new thing from nothing and give up their creative mind.

Actually, what is claimed to be a "new product" is a combination of two known products.

The discovery of the concept and formulation of gravity by Isac Newton is often told by the story of a falling apple from apple tree and planetary motion around the sun. Newton combined two existing observational facts: a falling apple and planetary motion. The relationship between them is far from what anyone imagined at that time.

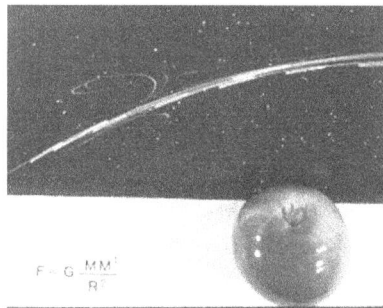

A hybrid car may be an example of innovation based on creativity. When Toyota faced an expensive improvement of car exhaust problem (reduction of the pollution, it costs a lot of money for just an improvement a few %), they combined two things -- engine and electric motor, reducing the problem by

50 %.

Hybrid car.

Take the example of a pencil. One end is round and soft, but the other end is sharp and hard. Two schools of scientists might argue for a long time. But it is often an open-minded third group that discovers that they are parts of pencil.

The following may not be a good example, but I understand there are two theories of the cause of the motion of tectonic plates: One is the mantle convection theory and the other is called "tablecloth cutting theory" (when you cut a table close in the middle, the two parts slide down from the table by themselves). Both seem to be right. There may be some a way to combine the two.

Charles Darwin was the most creative scientist in the history of science, because he combined all the living creatures in his evolution tree. However, there is a much simpler example of his creativity. When he was visiting the Indian Ocean, he noticed three kinds of atolls and classified them in order and suggested that they evolved as the sea level increased; this was later confirmed by drilling at an atoll. I spent a few months at an atoll in the South Pacific Ocean and went around other atolls, but did not recognize even there are different types. Darwin recognized three types of atolls and further synthesized them in terms of the sea level increase.

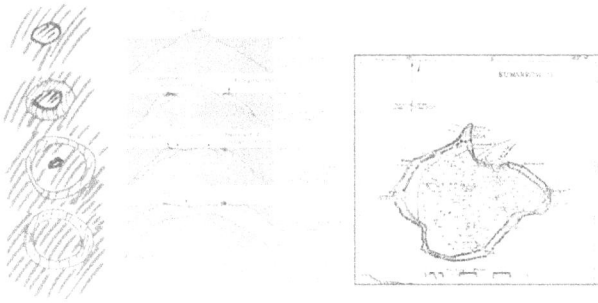

My humble examples

1. Geomagnetic storms

Textbook— "*Geomagnetism*" by S. Chapman and J. Bartels, Oxford Univ. Press, 1945.

The book had been the most authoritative book on geomagnetic storms for a few decades. It is mentioned that geomagnetic storms are caused by an impact of solar gas, which is marked by a sudden increase of the magnetic field of the earth, called by storm sudden commencement (ssc). The storm sudden commencement is followed by a large decrease of the magnetic field, called the *main phase;* such storms are the standard storm. The solar gas consisted of only protons and electrons.

Same as Figure 2.7.

Sydney Chapman suggested that I might study how the main phase occurs. He told me that he and many others had worked on the problem theoretically for 30 years, but could not solve the problem.

I thought that I should examine first magnetic records of many storms (I did not know much about geomagnetic storms) and found that geomagnetic storms develop in a great variety of ways, in addition to the standard way. I

285

was very confused and considered how they can be organized. I found that the storm sudden commencement does not indicate the beginning of storms.

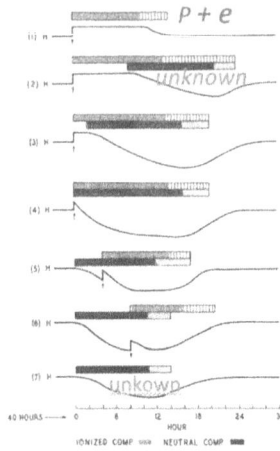

Same as Figure 2.13 (b).

When I finally classified and organized magnetic records in the way shown in the figure, I concluded that there must be a "unknown" factor (in addition to a flow of protons and electrons), which can cause the main phase.

When I tried to explain my finding in a scientific meeting: "You are not qualified to study geomagnetic storm by not knowing that the solar wind consists of protons and electrons." as has been taught in *Geomagnetism.*

A few years later, it was found that the "unknown" factor is solar magnetic field (IMF [-Bz] component).

2.   *Sunspots*

G. H. Hale was the first to recognize that sunspots have magnetic field (1908), which consists of a pair of positive and negative (or north and south) fields.

In 1961, H. W. Babcock theorized that a magnetic tube rises up from below the photosphere, and that its two cross-sections are a pair of spots. There is no way to confirm his theory, because there is no way to observe the tube below the photosphere. However, this is what has been basically believed even today (after 60 years).

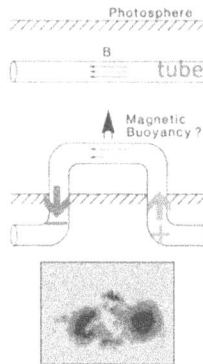

Same as Figure 7.4.

When I was scanning films of sunspots at the Kitt Peak Solar Observatory, I was surprised by the variety of appearance of sunspots. One of them is single spots. Since we are taught that sunspots consist of a pair (N/S or P/N) of spots in every textbook on the sun, I was surprised by their presence. In fact, there are many of them. Obviously, single spots are an 'odd' observed fact. I tried to find papers or recent books on single spots, but could not find about them. Finaly, I found them in the first (old) comprehensive book (before Babcock's theory was published) on the sun by Abetti (1955); he described them as solitary or independent spots. This confirmed that my finding is correct.

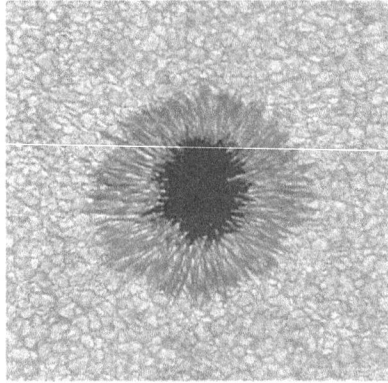

Same as Figure 7.48.

I asked about single spots to several solar physicists, including H. Zirin at Cal Tech and K. Shibata at Kyoto University. The best answer was a "broken pipe." The reason why I could not find any paper on single spot is perhaps that since a magnet should consists of N and *S (P and N), single spots (unipolar spots) should not exist. Thus, single spots are an 'odd' case and were avoided to be studied.*

Nevertheless, they *do exist*. Thus, I assembled many observed facts. One example is that there are weak positive and negative fields, and positive single spots are born in a positive weak field (vice versa). A pair of spots appears only at the boundary of positive/negative weak field, not in the middle of the weak fields. I reported these observed facts in Geophysical Research Letters.

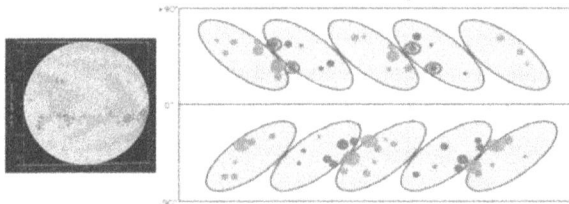

Same as Figure 7.49.

Further, I found a spot like a typhoon and found also many papers, reporting an upward flow of plasma from the top, downward flow outside conversing flow at the bottom, so that I thought that there is a cyclonic flow around a

288

sunspot.

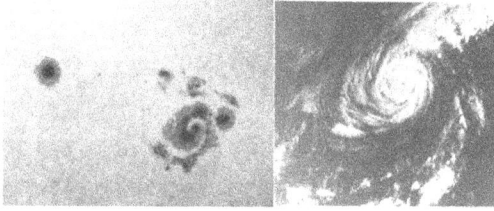

Same as Figure 7.62.

Based on these findings, I was convinced that a single spot is the unit of spots, not a pair of sunspots and that there was a cyclonic flow of plasma.

However, the reviewer of my paper stated "noble, but is not acceptable---". My study deviated too far from the common belief (Babcock's model).

I do not know the artist of the painting below. Suppose that it is Picasso's painting of Mona Lisa; if so, it might cost $100 millions. However, from Mona Lisa paradigm point of view, it is inaccurate and will be rejected, regardless how valuable it is.

Looking back

Most step-wise advances are seriously questioned when they emerge. This is simply due to the fact that since you are bound to be under a specific paradigm. You are not supposed to consider another paradigm (other than Garfield puzzle). As I said earlier, you need courage to pursue your idea.

Scientific methodology

As far as Earth sciences and space physics (including solar physics) are concerned, I believe that it is best to follow the order shown in the figure below. The situation may be different in other fields, such as elemental particle field; new theoretical predictions come first and are confirmed by large-scale experiments. Gravity was conceived by Newton and further by Einstein, but it is now tested by detecting gravity wave by astronomers.

In Earth sciences, such a theory before observation may cause confusion. In space physics, magnetic reconnection was presumed as early as 1958 and has dominated both auroral and solar physics. Although many satellites were sent to where it should occur, and no conclusive evidence was found. The author who is an expert in magnetic reconnection stated in his latest review paper that it is "mysterious" and "illusive" even after 60 years when it was presumed. It seems to be like a phantom theory, but is still believed by many.

8.1 Introduction

There are generally three ways to study in geoscience and natural science: (1) Observation (including analysis), (2) Morphological (synthesis) study and (3) Theoretical study. In general, most researchers in geosciences and natural sciences (including space physics) tend to specialize in one (1) or (3) or both.

Morphological or synthetic study is not often adopted.

One of the purposes of this chapter is to promote morphological study. This is because I believe that a good theoretical study should be based on a solid morphological study in natural sciences.

In geoscience and natural science, a new field is often birthed with an "*unexpected*" observation. I call it **Stage A**. **Stage A** is followed by **Stage B**: many observations to prove (or disprove) the unexpected observation or unthinkable idea. At this stage, a theoretical study may be confusing.

After **Stage B,** researchers try to assemble, synthesize or unify many observational facts (morphological study). After those early stages, theoretical models are proposed. This stage of synthesis is **Stage C.**

In **Stage D,** one of theoretical models becomes popular. Thus, practically, most researchers work under the popular one, establishing a **paradigm.** Thus, any ideas different from the prevailing paradigm are not welcomed and are often rejected by the paradigm participants.

Actually, many fields have repeated above sequence of Stages a few times in the past during their development. Many paradigms were established and gone in the past, so that we have already a large number of observed facts and a few syntheses (morphological studies) had already been made in most of the established fields in the past (Kuhn, 1962).

However, a synthesis effort of assembling observed facts is not often adopted in an established paradigm. Actually, this is one of the important steps in making a step-wise advance. This step will tell what observation or theory is needed. This can promote further advance of the field.

In space physics, an example of well-known syntheses is the morphological theory by Axford and Hines (1961), which was followed by many theoretical and observational studies. It may be interesting to note that they got their idea from Chapman's SD current in the ionosphere (Chapman and Bartels, 1950), which was dominated in the field. Thus, in Geomagnetism, there had been at least two paradigm changes in the past.

8.2 Assembling observed facts and recognize 'odd' facts

If one wants to make a step-wise advance in contributing to his/her field, what one can do?

One strategy is *assembling* all available observed facts and synthesizing them. It is in this process, in which one *might notice* an 'odd' piece, which is not consistent with an accepted theory.

There should be at least several 'odd' facts. However, it is not a good idea to look for the first 'odd' fact. It should be *unexpectedly* recognized during the assembling observed facts (synthesizing).

During assembling data, it is important to include ignored or dismissed data in the past (such as spotless solar flares or single spots (Chapter 7), which may have a new hint for future research. If such an 'odd' fact under a prevailing paradigm is found, it is crucially most important in advancing his/her field. What people ignored may be most important. The degree of importance of 'odd' pieces depends partly on the researcher's creativity, because the importance depends on if he/her can develop it into a new paradigm.

Then, there are three choices. The first is throw the 'odd' fact away, because it does not agree with his/her paradigm, and the second is to continue to find its location it fits. The third choice is to look for other 'odd' facts. If one can find one such an 'odd' fact, one can find several more, since natural phenomena are so complex. It is very important to find at least a few more 'odd' facts. This could indicate that the 'odd' facts are genuine, namely they are no longer 'odd'.

This chapter is prepared for those who choose the third choice.

The next step is to try to find a way to organize systematically the data *by completely putting aside the well-established theory or prevailing paradigm.* I define this step as synthesis, very often resynthesis, because in an advance field, it has been done at least once.

One of the reasons why a paradigm tends to stagnate is that contradicting observed facts are often disregarded or dismissed; such an 'odd' piece is thrown out because it is an obstacle in the prevailing paradigm.

8.3 Synthesis: Creativity is essential

Now, you have assembled a large number of observed facts with several odd pieces that do not agree with the prevailing paradigm.

Creativity in synthesis

It is in this step, in which creativity plays a crucial role.

Creativity is the formulation of a new idea by *bringing together at least two distinctly different observe facts*. It is not to produce something from nothing (as is often misunderstood).

Based on my experience, there are a few points to consider in conducting synthesis or resynthesis.

8.4 A few hints for synthesis or resynthesis

(a) Choose the simplest possible observational fact

In starting synthesis or resynthesis, my experience is that it is best to choose *the simplest possible* observational fact among all odd facts as the first step. One tends to be attracted by fascinating and complicated facts; the simplest observational facts are often dismissed (not worth studying or reporting).

(b) Synthesis is often criticized as "cherry picking."

Synthesis work is often criticized as cherry picking. Creativity is to bring many 'odd' pieces together and brings up a new idea. Such a criticism can be defended if one has a *basic line of thought in the synthesis process*. I took the electric current approach throughout my study of auroral substorms and solar flares, instead of the prevailing the magnetic field line approach. In trying to synthesize all the facts, a new idea might come up (creativity).

(c) Synthesis should be as quantitative as possible

Synthesis or resynthesis should be as quantitative as possible by determining physical quantities, such as current intensity (A), watt (W) and Joule (J). This is because its main purpose is to build a quantitative morphological theory, which can be handed over to theorists in developing their theoretical models.

(d) Unthinkable idea

Unthinkable idea should be based on solid data analysis, not by tossing it up. Even so, it can so easily be criticized. When it is criticized, one way to overcome criticism is to consider that the criticism may be like criticizing a baby by considering the baby is an adult. This is because unthinkable idea is just like a new baby.

Thus, one must convince oneself first its soundness by reexamining a large assembled observed facts and by making a simple quantitative check.

(e) Summary

In geoscience and natural science, I believe that *any worthwhile mathematical theory must be based on a solid morphological theory,* which can be built on the basis of a large number of solid observed facts. Mathematical theories without a solid morphological study (without solid observed facts) are simply confusing. This is certainly the case in geoscience and natural science (somewhat different from physics). In space physics, such an example is magnetic reconnection. Many theorists tried to consider this phantom theory.

A solid morphological study can be developed *into a solid mathematical theory by theorists.*

Theorists can also quantitatively and logically scrutinize morphological theory and improve it. This is certainly one way to advance a scientific field.

Reference

Axford, W. I. and Hines, C. O., 1961, A unifying theory of high-latitude geophysical phenomena and geomagnetic storms, Can. J. Phys., **39**, 1433.

Chapman, S. and Bartels, J., 1950, *Geomagnetism*, I & II, Oxford Univ. Press.

Kuhn, T. S.,1962, *The Structure of Scientific Revolutions* (2nd ed), Univ. Chicago Press.

Name Index

Berkner, L.: 29

Bertels, J.:28,34,63,246

Bharti, C.: 259,267

Borovsky, J. E.: 193,236

Brekke, A.: 138

Bristow, W. A.: 133

Biermann, L.: 25

Birkeland, K. R.: 17,31,85

Borovsky, J. E.: 193,236

Bostrom, R.: 132,140,146,159,161,206

Bramhall, E. H.: 90

Bristow, W. A.:133

Buchau.Y.: 73 74,78

Bulraga, L. F.: 215,222

Burch, J. L.:213

C

Cahill, L. J.: 50,53

Cain, J. C.: 52,60

Carlheim-Gyllenskold, N.: 29

Cassak, P. A.: 194,214

Donovan, E.: 181

Dodson, H.:197

Dryer, M.: 217

Dungey, J. W.: 31,49,50,51,122

E

Eather, B.: 99,106

Eddington, A.: 65

Egeland, A: 173

Elvey, C. T.: 30,71,100

Emperor Showa: 83

F

Fairfield, D. H.: 50

Feldstein, Y. I.: 72,97,109,114,123

Ferraro, V. C. A.: 20,31,40,52

Feynman, Jo Ann: 114

Feynman, R.:113

Fletcher, L.: 212

Merritt, R. P.: 144

Min, S.: 198,265

Moltke, H.: 5

Moore, R. I.: 207, 212

Mozer, F. S.: 168

N

Nagata, T: 37,85

Nansen, F.: 4

Nakagawa, Y.: 256

Neugebauer, M.: 49

Ness, N. F.: 60,213

Newell, P. T.: 168

Nikolsky, A.P.:189,190

Norlund, A.: 259

O

Olson, W. P.: 128,152

Onwumechili, A.: 85

Osaka, J.: 36

Subject Index

C

Current-carrying electron

D

Deflation of the inner magnetosphere: 146, 157

Diffuse aurora: 81

Dipolarization: 164

Directly driven (DD) current: 132

Disparition brusque (DB): 203

Discovery of O^+ ions: 56

Discovery of the Van Allen radiation belts: 26

Double layer

Double layer (aurora): 146,167

Double layer (solar flare): 202,207

Double layer (corona): 234

Dst index: 41

Dynamo (Power supply)

Dynamo (aurora): 121

Dynamo process (solar flare): 200

Dynamo process (coronal): 235

Dynamo process (solar wind): 239

Dungey's suggestion on the IMF (-Bz): 49

E

Electric current circuit

Energy accumulation:

Expansion phase

Interplanetary shock wave: 43,217,224

Inflation of the inner magnetosphere:145, 152

Ionization of the corona: 233

Ionospheric substorm: 106

Ionospheric electric current during substorms:131,132

Isolated spot:255

Longitudinal distribution of solar wind speed: 241

M

Magnetic arcade: 199

Magnetic buoyancy: 254

Magnetic cloud (MC): 16,41,51,78,216,221,222 217

Magnetic field line approach: 24

Magnetic field of the heliosphere: 240

Magnetic loop (solar flares): 203

Magnetic loop (corona): 236

Magnetic reconnection: 211

Magnetic shear: 204

Magnetohydrodynamics (MHD): 22,23,163,167

Magnetopause: 122

Magnetosphere: 21

Magnetosphere-ionosphere coupling system: 217,132

Magnetospheric substorm: 106

Magnetic equator (source surface): 243

Magnetic field line approach: 24

Magnetic monopole: 263

Magnetic reconnection: 211

Main phase of geomagnetic storms: 40,42

Mantle flow: 59

Maunder's observation: 16

Magnetosphere-ionosphere (M-I) coupling system)

Primary M-I coupling system: 127

Secondar M-I coupling system: 161

Micropulsations: 36

N

NASA Galileo flight: 99,101

Neon sign: 120

O

Omega band: 92,95, 180

Onset

Growth phase: 126, 149

Expansion phase: 93,156,162

Open magnetosphere:77

Open region:77

Oxygen (O^+) atom: 56,167,183

P

Pair of spots: 268

Patches (aurora): 92,95,179

Pedersen current: 165,167

Phenomenological study:

Phenomenological study (auroral substorms): 89

Phenomenological study (solar flares):197

Phenomenological study (sunspots): 252
Photospheric dynamo: 199

Plasma flow around a single spot: 264

Polar magnetic substorm: 57

Polar cap index: 126

Polar magnetic changes during the extremely quiet days: 141

Poleward expansion (advance) of auroras: 92,95,163,174

Pore: 255, 260

Power of dynamo

Power of dynamo (aurora):121

Power of dynamo: (solar flare): 199 :

Poynting flux: 152

Prediction of geomagnetic storms: 223,246

Prediction of solar flares: 205

Primary M-I coupling system:127,152,169,171

Prominence: 206

Propagation of shock wave: 217,220,224

Proton aurora substorm: 106

R

Radio absorption substrom: 106

Radio star scintillation: 223

Ray structure of auroral curtain: 8

Recovery phase: 171

Recurrent geomagnetic storm: 16,246,256

Red aurora: 1

Ring current

Ring current: 34,42, 52,55,56,79,80,182

Magnetic field of the ring current: 54

Ring current particle: 56,153,170

Formation of the ring current: 54

Ring current and substorm relationship: 57,170

S

Saturn's aurora: 186

Spotless flare: 197

Storm sudden commencement (ssc): 40

Stream: 16, 247

Streamer: 249

Sudden brightening of arc:43,92,93,217,224

Sudden impulse (si): 95

Sunspot: 252

SuperDARN radar: 134

T

Torch: 180

Torsional oscillation of the sun: 267

Twenty seven-day- recurrent tendency: 16,246

U

UF flow: 241

UL current: 132,136,148,156

Ul-Dst relationship: 184

Unipolar induction system 239:

Unipolar magnetic region: 257

Unipolar spot: 255

Unloading (UL) current: 182

"Unknown" factor: 49, 125

Ulysses observation of the solar wind: 241

V

Van Allen's discovery of the radiation belts: 26

Van Allen belt: 99

VLF substorm: 106

W

Westward traveling surge: 92,95, 177

X

X-line: 213

X-ray substorm:106

www.ingramcontent.com/pod-product-compliance
Lightning Source LLC
Chambersburg PA
CBHW021029210326
41598CB00016B/951